Disaster Preparedness and Climate Change in Cuba

Lexington Studies on Cuba

Series Editors: John M. Kirk, Dalhousie University, and Mervyn Bain, University of Aberdeen

This series will publish texts on all aspects of Cuba, focusing on the post-1959 period. It seeks to be truly interdisciplinary, with studies of all aspects of contemporary Cuba—from foreign policy to culture, sociology, to economics. The series is particularly interested in broad, comprehensive topics (such as women in Cuba, economic challenges, human rights, the role of the media, etc.). All ideological positions are welcomed, with solid academic quality being the defining criterion. In exceptional circumstances edited collections will be considered, but the main focus is on high quality, original, and provocative monographs and innovative scholarship.

Recent Titles in This Series

Disaster Preparedness and Climate Change in Cuba: Management and Adaptation
Edited by Emily J. Kirk, Isabel Story, and Anna Clayfield
Cuban International Relations at 60; Reflections on Global Connections
Edited by Mervyn J. Bain and Chris Walker
Soviet Influence on Cuban Culture, 1961–1987: When the Soviets Came to Stay
By Isabel Story
Entangled Terrains: Empire, Identity, and Memory of Guantánamo
By Asa McKercher and Catherine Krull
God and the Nation: A Social History of Cuba's Protestants
By James A. Baer
The People's Professors: How Cuba Achieved Education for All
By Kathryn Moody
Youth and the Cuban Revolution: Youth Culture and Politics in 1960s Cuba
By Anne Luke
Cuba's Forgotten Decade: How the 1970s Shaped the Revolution
Edited by Emily J. Kirk, Anna Clayfield, and Isabel Story

Disaster Preparedness and Climate Change in Cuba

Management and Adaptation

Edited by
Emily J. Kirk, Isabel Story, and
Anna Clayfield

LEXINGTON BOOKS
Lanham • Boulder • New York • London

Published by Lexington Books
An imprint of The Rowman & Littlefield Publishing Group, Inc.
4501 Forbes Boulevard, Suite 200, Lanham, Maryland 20706
www.rowman.com

6 Tinworth Street, London SE11 5AL, United Kingdom

British Library Cataloguing in Publication Information Available

Library of Congress Cataloging-in-Publication Data

Names: Kirk, Emily J., editor. | Story, Isabel, 1987- editor. | Clayfield, Anna, 1983- editor.
Title: Disaster preparedness and climate change in Cuba : management and adaptation / edited by Emily J. Kirk, Isabel Story, and Anna Clayfield.
Description: Lanham : Lexington Books, [2021] | Series: Lexington studies on Cuba | Includes bibliographical references and index. | Summary: "Disaster Preparedness and Climate Change in Cuba discusses how, over six decades, Cuba developed a world-leading model of disaster management and climate change adaptation. Comprised of leading scholars and policy makers in the field, this volume questions what makes Cuba's effective model so distinctive and what others can learn from it"— Provided by publisher.
Identifiers: LCCN 2021025733 (print) | LCCN 2021025734 (ebook) | ISBN 9781793651310 (cloth) | ISBN 9781793651327 (epub)
Subjects: LCSH: Emergency management—Cuba. | Climatic changes—Cuba.
Classification: LCC HV551.5.C9 D57 2021 (print) | LCC HV551.5.C9 (ebook) | DDC 363.34/525097291—dc23
LC record available at https://lccn.loc.gov/2021025733
LC ebook record available at https://lccn.loc.gov/2021025734

Contents

List of Acronyms

ACC	Cuban Academy of Sciences (Academia de Ciencias de Cuba)
ANAP	National Association of Small Farmers (Asociación Nacional de Agricultores Pequeños)
BPD	Production and Defense Brigades (Brigadas de Producción y Defensa)
CCCCC	Center for Climate Change for the Caribbean Community
CDR	Committees for the Defense of the Revolution (Comités de Defensa de la Revolución)
CITMA	Ministry of Science, Technology, and Environment (Ministerio de Ciencias, Tecnología y Medio Ambiente)
CLAMED	Latin American Center for Disaster Medicine (Centro Latinoamericano de Medicina del Desastre)
COMARNA	National Commission for the Protection of the Environment and the Rational Use of Natural Resources (Comisión Nacional de Protección del Medio Ambiente y los Recursos Naturales)
COMECON	Council for Mutual Economic Assistance (CMEA)
CMEA	Council of Mutual Economic Assistance (COMECON)
CNE	National Energy Commission (Comisión Nacional de Energía)
CTC	Workers Central Union of Cuba (Central de Trabajadores de Cuba)
DC	Civil Defense (Defensa Civil)
DCN	National Civil Defense (Defensa Civil Nacional) (EMNDC)
EDF	Environmental Defense Fund
EJT	Youth Labor Army (Ejército Juvenil de Trabajo)

ELAM	Latin American School of Medicine (Escuela Latinoamericana de Medicina)
EMNDC	National Civil Defense (NDC) (Estado Mayor Nacional de la Defensa Civil)
FAO	Food and Agricultural Organization
FAR	Revolutionary Armed Forces (Fuerzas Armadas Revolucionarias)
FMC	Federation of Cuban Women (Federación de Mujeres Cubanas)
INSMET	Institute of Meteorology (Instituto de Meteorología)
IPF	Physical Planning Institute (Instituto de Planificación Física)
ISDR	International Strategy for Disaster Reduction
JUCEI	Coordination, Implementation and Inspection Boards (Juntas de Coordinación, Ejecución e Inspección)
MEP	Ministry of the Economy and Planning (Ministerio de la Economía y Planificación)
MES	Ministry of Higher Education (Ministerio de Educación Superior) (MES)
MFP	Ministry of Finance and Prices (Ministerio de Finanzas y Precios)
MINAG	Ministry of Agriculture (Ministerio de Agricultura)
MINAL	Ministry of Food Industry (Ministerio de Industria Alimentaria)
MINED	Ministry of Education (Ministerio de Educación)
MINEM	Ministry of Energy and Mines (Ministerio de Energía y Minas)
MINFAR	Ministry of the Revolutionary Armed Forces (Ministerio de las Fuerzas Armadas Revolucionarias)
MININT	Ministry of the Interior (Ministerio del Interior)
MINSAP	Ministry of Public Health (Ministerio de Salud Pública)
MNR	National Revolutionary Militias (Milicias Nacionales Revolucionarias)
MTT	Territorial Militia Troops (Milicias de Tropas Territoriales)
OMI	Military Organization of Industries (Organización Militar Industrial)
PCC	Cuban Communist Party (Partido Comunista de Cuba)
PIS	Comprehensive Health Program (Programa Integral de Salud)
RRMC	Risk Reduction Management Center
SDPE	System Management and Planning of the Economy (Sistema de Dirección y Planificación de la Economía)
UNDP	United Nations Development Programme
USSR	Union of Soviet Socialist Republics

Acknowledgments

This project was funded in part by the Canadian Institutes of Health Research (CIHR), Ford Foundation, Reynolds Foundation, and Social Sciences and Humanities Research Council (SSHRC).

Introduction

The early years of the Revolution were characterized by a succession of challenges and numerous, successful mass mobilization drives to confront them. The latter included, among others, the formation of the National Revolutionary Militias (Milicias Nacionales Revolucionarias, or MNR), the 1961 Literacy Campaign, the creation of the Committees for the Defense of the Revolution (Comités de Defensa de la Revolución, or CDR), and the establishment of the Popular Defense. After centuries of Spanish colonial rule, followed immediately by decades of U.S. neo-colonialism, and later the successful rebellion against the Batista dictatorship, the nascent revolutionary leadership was focused on protecting and defending the hard-won Revolution. Key to defensive strategies was the use of popular participation and mobilization. Without doubt, "defense" in the broadest possible sense was understood as integral to the success of the revolutionary project and heavily emphasized within its formative first years.

However, despite the implementation of various national defense strategies, the revolutionary leadership and the Cuban population were unprepared for confronting the extensive damage wrought by natural disasters. While the island could already count on a force of Red Cross workers and firemen charged with responding to disasters, when Hurricane Flora hit Cuba in 1963, it became clear that the existing measures were profoundly inadequate in this respect. Some 1,200 Cubans were killed and several hundred thousand were left homeless. Flora was "one of the worst natural disasters in Cuba in modern times, but also the first one that the revolutionary government had had to face" (Alonso and Palmarola 2014, 118). Despite its relatively limited strength—it was categorized as a level 2 on the Saffir–Simpson Hurricane Scale—Flora had a devastating effect not just on Cuba but also on the Caribbean region more generally, killing between 7,000 and 8,000 people (Schwartz 2015). Bar the Great Hurricane of 1780, Flora was (and remains) the deadliest hurricane in the Caribbean's history. Given the profound destruction and lives lost, Flora has remained deeply engrained within Cuba's cultural memory. Indeed, it has since been widely regarded as the decisive event that re-directed the revolutionary leadership's view on defense to include protection against, and preparedness for, natural hazards. Along with the other challenges threatening its survival and success, disaster management and risk reduction thereafter became embedded into the core values and objectives of the Revolution.

Since Flora, Cuba's leadership and population have continued to focus their efforts on managing natural disasters, with the view to minimizing lives lost and protecting infrastructure and the economy. Like the revolutionary project itself, the island's disaster management system has continued to evolve, in line with the emergence of new challenges, the availability of resources, and the needs of the population. For example, the Popular Defense, founded in 1962, was re-established as the Civil Defense in 1966; the Institute of Meteorology (Instituto de Meteorología, or INSMET) was founded in 1965; USSR–Cuba relations started to include collaborative research on extreme meteorological events in the 1970s and 1980s; the annual national practice drill, *Meteoro*, was instituted in 1986; in 1992, Fidel Castro announced to the world at the UN's "Earth Summit" that environmental protection was crucial and would be more fully incorporated into Cuban development plans; the island launched its national "Energy Revolution" in 2005; *Tarea Vida* (Project Life), Cuba's comprehensive strategy to combat climate change, was ratified in 2017; and the 2019 Constitution was again broadened in scope to include further legal protection and support for climate change adaptation initiatives and environmental protection.

While Cuba's approach to disaster preparedness and management was originally focused on prevention and risk reduction, it has evolved over the decades to place an increasing emphasis on the impacts of climate change. It first began by stressing the value of environmental protection and conservation and continued to broaden its scope by incorporating climate change and the need for mitigation and adaptation strategies. This shift has occurred largely as a result of Cuba's geographic vulnerability to climate change and its status as a middle-income country. Research suggests that without an action plan to mitigate the predicted 1-meter rise in sea level, by 2050, 122 Cuban coastal towns and nearby agricultural lands will be destroyed and water supplies adversely affected. An estimated 2.3 percent of the island would ultimately be submerged. Climate change mitigation, along with disaster preparedness, strategies are thus key to the island's very survival, not just that of the Revolution.

For decades, Cuba's disaster preparedness and management model has been lauded internationally, including by the World Health Organization and the United Nations. Nearly twenty years ago, the International Federation of the Red Cross (IFRC) 2002 Disaster Report stated that Cuba "gives us a model of effective government-driven disaster preparedness" (Thompson and Gaviria 2004, 6). Similarly, the United Nations Office for Disaster Risk Reduction described Cuba as "a model in hurricane risk management" (UNDRR 2004). In 2006, the United Nations Emergency Relief Coordinator also noted that "Cuba is one of the best prepared, if not the best prepared for natural disasters. . . . The same hurricane which would take zero lives in Cuba would kill massively in Haiti" (Otto 2006). Similar assessments continue to be articulated to date.

The success of Cuba's approach can perhaps best be illustrated by the regional figures of lives lost during times of disasters. When Hurricane Matthew struck the Caribbean region in 2016, for example, zero deaths were reported in Cuba, while the hurricane caused forty deaths in neighboring United States and over 1,000 fatalities in Haiti. Similarly, in the month of September 2017, when two Category 5 hurricanes lashed the Caribbean (one, Hurricane Maria, centered on Puerto Rico and the United States, and the other, Hurricane Irma, on Cuba) there were notable, and arguably extreme, differences in the numbers of lives lost. Hurricane Maria caused a reported 66 deaths in the United States and at least 2,975 deaths in Puerto Rico, yet only 10 deaths resulted from Hurricane Irma when it made landfall in Cuba. The figures become more staggering when one considers that Irma caused $13 billion in damage and led to the evacuation of 2 million people on the island. The consistently low numbers of fatalities during and following a major disaster can be explained by Cuba's distinctive and comprehensive approach to disaster preparedness and management.

Despite this exceptional success rate regarding minimal loss of life and property, research on Cuba's approach to disaster management has nonetheless been limited. This lack of investigation is largely down to the system's breadth, complexity, and varied foci. It comprises multiple layers and components, including the Civil Defense and Armed Forces systems, electorate and national political systems, international relations, marine science, ecology, agriculture, law, healthcare, education, development, and ideology, among others. The purpose of this volume is to explore, untangle, and illuminate these overlapping and idiosyncratic components of Cuba's approach to disaster management and climate change adaptation by drawing on the findings of a variety of noted specialists from distinctive fields. In so doing, the book seeks to offer explanations as to how and why Cuba's specific strategy has proven to be so effective.

Sub-topics are organized into fourteen chapters, each focusing on a different, complementary, component. Beginning by focusing on the evolution and distinguishing qualities of the island's approach, it subsequently includes an analysis of climate change management and specifically the adaptation strategies employed. Emily J. Kirk's opening chapter outlines the development of Cuba's disaster preparedness and risk reduction model, while also highlighting its uniquely Cuban characteristics. Jessica Hirtle's contribution then assesses the ways in which the system operates within a structure of "formal," "semi-formal," and "informal" strategies. The importance of "defense" as one of the Revolution's underpinning values is then explored in Hal P. Klepak's chapter on the evolution and role of the Armed Forces and Civil Defense. Exploring related themes, Isabel Story traces the development of Cuban–Soviet relations and cooperation in the field of disaster preparedness and management, linking developments in research collaboration back to wider trends and tensions in this historic alliance. Anna Clayfield's chapter on the Bastión and the shaping of a *Pueblo Combatiente* examines one of the most prominent defense initiatives of the last six decades, "The War of All the People" (*Guerra de Todo el Pueblo*), launched in response to a re-heating of Cold War tensions. The book then turns its attention to the two pillars of the Revolution: education and health care. Emily J. Kirk's second contribution investigates the impact of education in post-1959 Cuba to ensuring an informed and highly prepared population. Specifically, the importance of *Meteoro*, a national two-day training exercise before the start of the hurricane season, is explored. Also on the theme of education, Guillermo Mesa Ridel's chapter summarizes how the Cuban health sector has developed to include disaster medicine, specifically from 1996 to 2004. In a similar vein, Lauren Collins' chapter demonstrates how the steady advances in education, combined with the evolution of a participatory local polity, have shaped Cuba's path to effective disaster management.

The focus of the volume then shifts toward the impact of climate change and environmental resilience. Helen Yaffe's chapter discusses *Tarea Vida* and reviews the island's earlier work on environmental protection and sustainable development. Valerie Miller's contribution assesses Cuba's role in marine conservation and the importance of collaboration across Cuba, the United States, and the wider region. The results of an international research team, led by Margarita Fernández and comprising noted specialists Leidy Casimiro Rodríguez, Luis L. Vázquez, and Giraldo Martín Martín, present empirical data from Cuba to highlight important elements of resilience in the country's food system. These include results from a socioecological analysis, a national project that integrates energy production with food production, and an agroecological co-innovation experience. Helen Yaffe's research on energy alternatives is subsequently featured, with a chapter that traces the development of alternative energy sources in revolutionary Cuba, focusing in part on the so-called "Energy Revolution." Taking a broad look at the importance of the island's legal framework, Shawn H. E. Harmon and

Emily J. Kirk's chapter centers specifically on the legal elements of Cuba's hazard preparedness, mitigation, management, and recovery systems. Finally, Jessica Hirtle offers a comprehensive timeline of significant events and their impact on Cuba's approach to the management of, and adaptation to, natural hazards and climate change.

It is important to note that, given the diversity of contributing authors from multiple disciplines, this volume will include some different vocabulary and wording in its descriptions and analyses of the subject at hand. A distinction between a "natural hazard" and "natural disaster" should be noted, for example: a "natural hazard" is defined as an unexpected and/or uncontrollable natural event of unusual magnitude that threatens health, life, property, and/or the environment and a "natural disaster" is a natural hazard that results in health, life, property, and/or environmental damage on a scale sufficient to require assistance. Similarly, and although their meaning is identical, the use of "cyclone" is more commonplace in languages other than English, particularly during the twentieth century, while "hurricane" has become increasingly prevalent in contemporary studies published in the English-speaking world.

It is also important to highlight that "global south" is not utilized throughout this volume as a strictly geographical reference. This study uses "global south" in place of other terms such as "developing," "underdeveloped," "Lower Income Country (LIC)," and "Third World" (Kirk and Walker 2020). As Kloß (2017, 1) explained: "The global south is not an entity that exists per se but has to be understood as something that is created, imagined, invented, maintained, and recreated by the ever-changing and never fixed status positions of social actors and institutions." The global north and "the West" both act as an important point of contrast in terms of "development."

Climate change and the devastation it is wreaking now constitutes an ever-growing focus of Cuba's disaster management planning and execution, and as such is writ large throughout this volume. Natural disasters and severe meteorological events are without doubt a by-product of climate change. Although the impacts of these disasters are felt across the world, they disproportionally impact countries of the global south and marginalized communities. Sims and Voglemann (2002) calculated that natural disasters more than quadrupled from 1952 to 2002 and caused approximately 1,000,000 deaths during the twentieth century. While they rarely cause loss of lives in Cuba, the island does face two to three significant natural disasters every year, usually in the form of hurricanes. As climate change develops, the intensity and frequency of natural disasters will only continue to grow, in Cuba and elsewhere. The Cuban approach, with its resulting low loss of life and high degree of infrastructure protection, offers important lessons in alternative strategies for disaster preparedness and management that ultimately make a meaningful contribution to efforts to address this global challenge.

BIBLIOGRAPHY

Alonso, Pedro Ignacio and Hugo Palmarola. 2014. *Panel*. London: Architectural Association Publications

Kirk, Emily J., and Chris Walker. 2020. "Meeting the UN's Development Targets in Cuba: Lessons from the Global South." *International Journal of Cuban Studies* 12 (2): 252–271.

Kloß, Sanah T. 2017. "The Global South as Subversive Practice: Challenges and Potentials of a Heuristic Concept." *The Global South* 11 (2): 1–17.

Otto, Lilja. 2006. "Cuba's Hurricane Resilience—Solidarity and Readiness." *Yes Magazine*. Accessed January 28, 2020. http://www.yesmagazine.org/issues/5000-years-of-empire/cuba2019s-hurricane-resilience.com.

Schwartz, S. 2015. *Sea of Storms: A History of Hurricanes in the Greater Caribbean from Columbus to Katrina*. Princeton, NJ: Princeton University Press.

Sims, Holly, and Kevin Vogelmann. 2002. "Popular Mobilization and Disaster Management in Cuba." *Public Administration and Development* 22: 389–400.

Thompson, Martha and Izaskun Gaviria. 2004. *Weathering the Storm: Lessons in Risk Reduction from Cuba*. Boston, MA: Oxfam America.

UNDRR. 2004. "Cuba: A Model in Hurricane Risk Management." United Nations Office for Disaster Reduction. September 14. Accessed October 21, 2020. https://www.un.org/press/en/2004/iha943.doc.htm.

Chapter 1

Disaster Preparedness and Management

What Makes the Cuban Approach Different?

Emily J. Kirk

Cuba has been lauded by, among others, the United Nations, World Health Organization, Red Cross, and Oxfam for its efforts regarding disaster preparation and management, distinguishing the Cuban approach from other global south countries. Since 1963, the Caribbean island nation has continued to develop, assess, and adapt its disaster management strategies, resulting in a distinctive approach to meet the challenges faced by natural hazards. Cuba's approach has evolved to include comprehensive strategies to protect its population, infrastructure, and the environment from a wide variety of natural and meteorological threats. Of particular note, Cuba's approach to disaster preparedness and management has been internationally recognized for its consistent results of noticeably low numbers of lives lost during major meteorological events, especially when compared to other neighboring countries. For example, when Hurricane Matthew hit in October 2016, over 1,000 fatalities were reported in Haiti and over one month following the disaster, 1.4 million Haitians remained in need of emergency assistance. By contrast, in Cuba, although severe floods were reported in the provinces of Granma, Holguín, and Santiago de Cuba, 1.3 million people were evacuated, and damage was calculated at over $2.5 billion, not one death was reported. To put this in further context, while zero deaths occurred in Cuba as a result of the Hurricane, forty deaths were reported in the United States (Kirk 2017).

The diverse experiences of Hurricane Maria and Hurricane Irma further illustrate the significant results of Cuba's distinctive approach. Hurricane Maria hit Puerto Rico in September 2017 as a Category 5 (the highest category on the Saffir–Simpson scale) and was widely considered to be one of the most intense natural disasters to hit the island in recorded history. The widespread devastation was highlighted in the death toll associated with the disaster. While the U.S. state authorities reported 66 deaths as a result of the hurricane (Guha-Sapir and Checchi 2018), researchers from George Washington University concluded that there had been some 2,975 deaths associated with the hurricane between September 2017 and February 2018. These deaths were both direct (such as drowning or being crushed by a falling building due to high winds) and indirect results (including contaminated water and spread of disease) of Maria. Furthermore, researchers concluded that elevated risk for citizens continued beyond the observation period of the study and ongoing deaths and health issues associated with the damage were expected (George Washington University 2018).

In the same month, a powerful Category 5 hurricane also hit Cuba, Hurricane Irma. Like Maria in Puerto Rico, it was record-breaking—becoming the first recorded hurricane to maintain Category 5 intensity for three consecutive days as it crossed Cuba. Like Maria, Hurricane Irma was extremely destructive, yet notably minimal loss of life occurred in Cuba. For example, while damages directly caused by the hurricane were calculated at over $13 billion, 158, 554 houses were damaged or destroyed, and some 2 million people were evacuated (of a population of 11.3 million), only ten deaths were reported (*Cubadebate* 2017). Indeed, in a 2002 Disaster Report, the International Federation of the Red Cross (IFRD) asked, "What is the secret of Cuba's success?" (Thompson and Gaviria 2004, 6). This chapter will illuminate the specific approach employed in Cuba, how it evolved, and what makes it unique.

THE EVOLUTION OF CUBA'S APPROACH

Cuba's contemporary system is rooted in the events of 1963, following Hurricane Flora, which devastated the island. While only rated as a Category 2 on the Saffir–Simpson scale, the hurricane caused significant damage and over 1,200 deaths. Evacuations and rescue attempts were haphazardly organized, while crops and houses suffered substantial damage.

Significantly, it is not uncommon when discussing disaster preparedness and risk reduction in Cuba for older generations to recount their experiences during and after Hurricane Flora. Emotional stories describe the damage in detail: houses being completely washed way; families losing all of their possessions; seeing human bodies, animal carcasses, and rotten food floating in the water; the terrible smell in the flooded areas; the subsequent struggle to find clean drinking water and food; and the communicable diseases and health issues that resulted from the stagnant and polluted water.[1]

While remembered for its devastation, Hurricane Flora is equally understood as the catalyst for a dramatically improved disaster preparedness system. Indeed, it has consistently remained in the cultural memory of the Cuban population and leadership as a time of significant damage and loss, as well as the major turning point for the island's approach to natural hazards. The significant losses suffered as a result of lack of preparation and management sparked the Cuban leadership's desire to establish and develop a comprehensive disaster preparedness and management plan, as attention shifted to how to save Cuban lives and property (Mesa 2008).

Following the chaos of Flora, problems encountered were thoroughly analyzed, as the leadership sought to improve the island's reaction and response to natural hazards (largely characterized at this time by floods and hurricanes). By 1965, the Institute of Meteorology (Instituto de Meteorología, or INSMET) was founded and given the responsibility of detecting, monitoring, and communicating hazards, such as hurricanes, high winds, and heavy rainfall (Castellanos Abella and Wisner 2019).

In addition, a major change instituted involved shifting the aims and responsibilities of the Popular Defense. Founded in 1962 with a view to maintain vigilance and protect the economic and political targets of the country, in 1966, it was re-organized as the National Civil Defense (Defensa Civil Nacional, or DCN) (also noted in documentation as Estado Mayor Nacional de la Defensa Civil, or by its Spanish acronym, EMNDC). While its predecessor was focused on military protection, the Defensa Civil (DC) was organized to protect the populations from both military threats and further natural disasters. This included organizing and implementing warning systems, plans for evacuations and shelter facilities, as well as managing safe drinking water, transportation, and communication (Puig González et al. 2010). From the late 1960s, the DC increasingly became the engine of Cuba's disaster preparedness and management approach. In essence, the establishment of the DC can be understood as the beginning of Cuba's approach, as it continued to evolve, adapt, and improve over time.

Throughout the 1970s and 1980s, the government continued to seek means of improving its disaster preparation approach. For example, this included seeking international cooperation plans to improve knowledge and strategies (such as with the USSR), establishing research initiatives, and developing an education system that includes comprehensive study of the sciences at all levels, including graduate and post-graduate training.

In 1976, and after careful analysis and planning, the organizational structure of the DC was significantly improved. While the leadership of the DC rested with the country's president, a strong and clear centralized organization format was instituted, with DC representatives at the national, provincial, and municipal levels of leadership. The aim was to improve the efficiency of organization and communication. At this time, the Civil Defense was also placed in charge of planning, organizing, and overseeing the protection, preparedness, response, and recovery of the Cuban population in times of disasters, particularly natural disasters (Defensa Civil 2017).

Education also became a progressively more important component of the preparedness and management system. For example, in 1986, a weekend-long training activity entitled *Meteoro* was instituted (Llanes Guerra 2008). Organized by the DC, the idea was for the population to actively practice, and participate in, disaster preparedness and response training. Specific drills were included, as detailed plans for evacuation and safety were practiced for a wide variety of possible disasters—from floods, to fires, and water contamination. Notably, participation has remained mandatory since *Meteoro*'s inception and includes both children and adults. The event has continued every year, in the month of May, as communities prepare for hurricane season.

Cuba faced significant challenges at the beginning of the 1990s as a result of the collapse of the Soviet Union, its largest trading partner, as well as the intensification of the United States' economic blockade (which, at the time, was going into its third decade). An intense economic crisis followed, known as the "Special Period." The first half of the 1990s in Cuba was thus characterized by extreme scarcities, including food, fuel, and economic allies. However, as a result, and in order to safeguard the revolutionary project, the country was required to seek alternatives to their now dated and struggling systems. For example, as a result of the fuel and food crises, emphasis was placed on developing sustainable and renewable energy sources. Agriculture practices too saw dramatic changes, as the industry was required to adapt to contemporary challenges, ultimately employing new agroecological methodologies and strategies. The Cuban population and government were again required to adapt to challenges, with the view to continue to protect the island with the resources available.

In addition, this period also prompted further changes within the DC system, including its central aims and the overall organization of the island's disaster preparedness model. Of particular note, in 1994, the DC and its role to protect the population and economy was reaffirmed and expanded. Moreover, funding was allocated to the DC to support their evolving preparedness plans and prevention measures. This is especially significant given the country's economic crisis and highlights the national dedication to the protection of the island's population, infrastructure, and economy. In terms of natural disasters, the management system was re-organized to include different stages or phases. These included "Normal," "Informative," "Alert," "Alarm," and "Recovery" (Pardo Guerra et al. 2017). Each phase was comprehensive in its planning and execution, as the DC, mass organizations, and government made the expectations and regulations clear for each of the stages.

For example, the Alert phase occurs forty-eight hours before a major meteorological event (usually hurricane) is scheduled to impact the island. It is characterized by the full mobilization of the population at risk. All students are sent home from schools, volunteers harvest vulnerable crops, agricultural equipment and infrastructure are protected, livestock are moved to assigned sheltered areas, and houses are secured. The state-run community shelters provided during this state are commonly schools or municipal buildings, prepared and maintained by the DCN. They are stocked with food, water, and medical supplies and are frequently inspected by infection control medical personnel to ensure they are safe and that the risk of communicable diseases is low. Similar to the Alert stage, each other phase is distinguished by specific mobilization and action plans. These stages have continued to be implemented, and expanded upon, depending on the given challenges and data available. In essence, and despite the crippled economy, the government continued to increase attention to disaster preparedness and management throughout the 1990s.

In 2004, Cuba was severely impacted by two major hurricanes within a relatively short period of time, Ivan and Charley. Following the destruction caused, it was determined that changes again needed to be made to the island's prevention and management system, as the Cuban approach continued to evolve. Notably, a process of analysis and improvement was undertaken. The following year, a new law was instituted, entitled Directive 1. The document outlined the changes that would be implemented, focusing primarily on prevention and risk reduction. For example, it included plans to incorporate disaster preparedness into all future socioeconomic planning, improve communication and knowledge dissemination, and required comprehensive risk assessments (República de Cuba, Consejo de Defensa Nacional 2005).

Directive 1 would become an impactful legal framework for the ongoing evolution of Cuba's system. In 2007, the Directive was updated to include the establishment of Risk Reduction Management Centers (RRMCs) for every provincial and municipal government with the view to increase research and decrease risks associated to specific regions (Pardo Guerra et al. 2017). The RRMCs have continued to play an important role in data collection, knowledge dissemination, and communication. The Directive was again updated in 2010, continuing its focus on risk reduction and preventive measures.

Similarly to the impact of hurricanes Ivan and Charley in 2004, the devastation caused by Hurricane Sandy in 2012 prompted the government to reassess its disaster preparedness and management strategy and include subsequent improvements. During Sandy, some 343,230 people were evacuated from the western province of Santiago de Cuba, $4.7 billion in damage was reported, 171, 380 houses were damaged, and 11 deaths were reported (Carrasco Martín 2012). As President Raúl Castro noted, the city of Santiago looked "like a city that has been bombed" (Galinsky 2012). As a result of the significant damage, the heads of various ministries including, among others, the Ministry of Energy and Mines (Ministerio de Energía y Minas, or MINEM), Ministry of Food Industry (Ministerio de Industria Alimentaria, or MINAL), Ministry of Agriculture (Ministerio de Agricultura, or MINAG), Ministry of Finance and Prices (Ministerio de Finanzas y Precios, or MFP), and Ministry of the Economy and Planning (Ministerio de la Economía y Planificación, or MEP) were stationed in the capital of Santiago de Cuba to help oversee the recovery and reconstruction.

The ongoing and severe impact of natural disasters, as well as the increasing effects of climate change across the island (such as flooding, drought, erosion, and water salinization), has resulted in the Cuban government's ongoing commitment to improve the island's approach to disaster preparedness and management. For example, among contemporary advancements to the system are the risk assessment studies (known by the Spanish acronym, PVR), largely directed by the Ministry of Science, Technology, and Environment (Ministerio de Ciencias, Technología y Medio Ambiente, or CITMA). The aim of these studies is to compile as much data possible on the specific hazards in each municipality (169 in total), in order to determine necessary improvements to protect lives, infrastructure, and the economy. While the studies originally focused on hurricanes (and by extension floods and high winds), they were expanded to include other forms of hazards, such as fires, drought, epidemiological crises, and landslides (Pardo Guerra et al. 2017). These PVR studies are updated annually in order to ensure the data is up-to-date and record any measurable changes due to climate change. These studies are considered to be vital for Cuba's preventive strategy, as they provide comprehensive assessments on potential problems and disasters, as well as how to manage them in order to minimize disasters.

Prompted by Flora, Cuba's system has evolved significantly from the early 1960s. Evolving over more than six decades, the Cuban approach to disaster preparedness and management has consistently demonstrated its efficacy as well as its ability to adapt and evolve. At its core, the island's system remains distinctly Cuban, with key characteristics that distinguish it from other approaches.

DISTINGUISHING CHARACTERISTICS OF CUBA'S DISASTER PREPAREDNESS AND MANAGEMENT SYSTEM

As illustrated throughout the evolution of the island's approach to disaster preparedness and management, there are several key characteristics that make it uniquely Cuban. These include Prevention; a Comprehensive National Civil Defense system; Education; Support of, and within, the legal system; Climate change recognition and adaptation; Involvement and participation of all Ministries; Solidarity; and Evolution.

Prevention: The main pillar of the island's approach is undoubtedly that of prevention. Similar to the view behind the country's healthcare system, the island's approach to disaster preparedness and risk reduction is rooted in the fundamental understanding that prevention saves lives. Moreover, it is widely believed that prevention is less costly than a strictly recovery-based approach. As such, it is considered paramount in the comprehensive strategy to protect the island from natural disasters, with the primary concern focused on the prevention of lives lost.

The significance placed on preventive measures is evident throughout the individual elements that make up the system and has consistently remained central to the approach's development. For example, prevention has been a key tenet of the legal structure that directs and promotes disaster management; included in the central themes within all education associated with disaster risk reduction and recovery; emphasized clearly throughout the media; and has remained among the research foci and aims of the island's extensive data collection and analysis strategies. Indeed, since the inception of Cuba's intricate approach to disaster mitigation, prevention has been a key theme, clearly present throughout the system's evolution and each of its individual components.

A Comprehensive National Civil Defense system: The Civil Defense system is complex, evolving since 1962 to establish and develop a comprehensive multidisciplinary network of national, provincial, municipal, and community leaders whose central focus is the protection of the population, particularly during times of disasters. It has developed over six decades to become a strong and centralized form of prevention, preparedness, response, and recovery from disasters. Organized from the highest level of leadership, the president, through to individual community leaders (such as hospital directors or school principals), the DCN has without question distinguished itself as a uniquely Cuban characteristic of the island's disaster management approach. As Castellanos Abella and Wisner (2019, 9) explain,

> The country is vertically organized into national, provincial, municipal, and popular council levels [. . .] A municipality has a "municipal council of defense" led by its president with a vice-president trained and responsible for day-to-day operations of civil defense. A province has a "provincial council of defense" which is also led by its president. In addition, there is a "national council of defense" led by the president of the Republic of Cuba.

This centralized model, present and active in each of Cuba's 169 municipalities, allows the country to easily implement and undertake both small-scale and large-scale preventative and protective strategies. In addition, the model is designed to facilitate data collection and research; this is particularly significant, for example, in times of disasters, when damage calculation and evacuation figures are crucial to response and recovery. In addition, the DCN's data is subsequently and consistently updated and analyzed, with the view to continue to improve the system's efficacy.

Education: Formal education (in schools, universities, state-sponsored print media, etc.) and informal education (media representations, cultural memory, etc.) have been extremely important in order to prepare Cuba's population for disasters. As one of the pillars of the Revolution, along with health care, education has consistently been understood as crucial to the county's development. In terms of formal education, literacy and attendance in primary and secondary schools were central to the revolutionary leadership's aims of the first decades of the Revolution. This served to establish and develop a widely literate and well-educated population (Smith 2018). Furthermore, children began learning about natural disasters and how to protect themselves and their families in primary school through secondary school, with graduate and postgraduate training either including information on, and centering around, natural hazards (Santos Pérez et al. 2018). For example, medical students are required to learn about the most common medical issues during and after a disaster, as well as related topics such as water safety and disease prevention.

Moreover, there are often public service announcements and information on the state-controlled TV and radio. Disaster preparedness education has also become commonplace in communities, as members of the DC and mass organizations continue to educate communities on best practices (such as, among others, evacuation plans, water quality assessments, risk reduction mapping, and energy conservation). Moreover, the annual weekend-long educative exercise, *Meteoro*, established over thirty years ago, requires all citizens to participate in drills to reinforce knowledge of where to go and what to do in the event of a given disaster (flood, fire, earthquake, etc.), as well as functions as a means of practical application of theoretical scenarios (such as rescue efforts). Safety information and education are top priorities during *Meteoro*, as the population is trained (and subsequently re-trained) on a wide variety of disaster-related information and protocols, which continue to develop and evolve based on updated data, research, and analysis. In regard to Cuba's overall approach, education has been a key tenet, illustrating the importance of an educated population to combat climate change and mitigate disasters.

Support of, and within, the Legal System: Cuba's approach is deeply engrained throughout the legal system, evolving to include updated legislation to support changes in the island's disaster preparedness strategies. The Cuban legal system contains four types of laws including, Laws, Decree Laws, Decrees, and Resolution; each type has been used to develop the country's legal structure, supporting efforts to mitigate disasters.

Beginning in 1962, several laws were enacted to protect the lives of Cubans in times of disaster. As of 2008, there were three laws, seven decree laws, thirteen decrees, twenty-one resolutions, and one directive that focused on disaster risk reduction as well as the protection of lives and wellbeing (Llanes Guerra 2008). Among the most significant laws was, and remains, Degree Law 170, which was ratified in 1997. It noted that the civil defense was responsible (using a system of prevention, preparedness, response, and recovery) for avoiding "human loss" and to "protect the population" as well as the economy and agriculture from natural hazards.

In addition, Cuba is one of the few countries with a constitution that outlines protection of the environment and the importance of disaster preparedness and management. The Cuban Constitution, which was instituted under the revolutionary government in 1976 (as well as amended throughout the years, most recently in 2019) expressly supports the protection of the population during times of disasters. For example, Article 96 of the initial 1976 version stated that the government was responsible for the "protection of lives and property in case of natural disasters" (República de Cuba 1976). Without doubt, the island's broad legal structure has translated into an inarguable foundation for the protective role of the government and the necessary compliance of the population in times of disasters, including evacuation, mobilization, and participation.

Recognition of, and adaptation to, climate change: While environmental protection has been important throughout the Revolution and was increasingly a topic of great concern to Fidel Castro, particularly since the 1990s, the emphasis on climate change has been a distinctive trait of Cuba's approach (Houck 2000). The island has consistently participated in global events, accords, and frameworks regarding environmental protection and, increasingly, those focused on climate change prevention and adaptation. These include the UN's Conference on Environment and Development, or "Earth Summit" (1992), Millennium Development Goals (MDGs) (2000), the Hyogo Framework (2005), Sendai Framework (2015), Sustainable Development Goals (SDGs), and the Paris Agreement (2016). For example, the Sendai Framework, developed by the UN Office for Disaster Risk Reduction, commits to: "The substantial reduction of disaster risk and losses in lives, livelihoods and health and in the economic, physical, social, cultural and environmental assets of persons, businesses, communities and countries" (UNISDR 2015).

Of particular note, Cuba has been increasingly focused on combating climate change due to the island's geography. As a result of Cuba's position in the Caribbean, it is at a measurably higher risk of the impacts of climate change. This includes, among others, loss of coastline due to rising sea levels, floods, water salinization, and soil erosion, as well as an increase in the frequency and intensity of hurricanes, rainfall, and high winds. For example, it was reported that increasing sea levels could cause the loss up to 5.45 percent of Cuba's coastal land by 2100, destroying numerous communities, water resources, and infrastructure (*Granma* 2017; Castellanos Abella and Wisner 2019).

In a national effort to adapt to the challenges faced by climate change, Cuba instituted *Tarea Vida* (Project Life) in 2017. Based on extensive research, data assessment, and numerous multidisciplinary studies, *Tarea Vida* was developed with the view to establish specific strategies to protect the Cuban population as well as help the island adapt to climate change. Plans and goals were identified in terms of importance and longevity and categorized as short term (2020), medium term (2030), long term (2050), and very long term (2100). With the development and implementation of *Tarea Vida*, Cuba is the only country in the world with a long-term state-led plan to protect the island (including the population, infrastructure, and environment) from climate change.

Involvement and participation of all ministries: Originally based on legislation that was instituted in 1966 (Law No. 1194), which focuses on the DC, all ministries in the Cuban government are required to participate in disaster preparedness and recovery (Defensa Civil 2017). In terms of a global comparison, this research has not found any evidence that other countries include a similar structure whereby every government ministry has clear, organized, documented, and previously agreed-upon roles and responsibilities within a comprehensive disaster preparedness and response strategy. It can thus be argued that Cuba is one of the only countries, if not the only country, that includes this attribute in their overarching disaster management approach. The significance of this is further demonstrated when compared to geographically close, global north countries. For example, in Canada in times of natural disasters, the primary body responsible for information and recovery is Environment and Climate Change Canada, the central department that works on all environment-related issues. In addition to their coordinating efforts, the armed forces are at times brought in to assist. Similarly, in the United States, while there is some support offered by Federal Emergency Management Agency (FEMA) and the armed forces, government departments are not coordinated or integral to the overall preparedness or recovery strategies.

Notably, in addition to the original 1966 law, further legislation and responsibilities continued to be included in Cuba, identifying the specific functions of all ministries; each ministry is tasked with specific responsibilities before, during, and after a disaster. For example, the Ministry of Tourism

(Ministerio de Turismo, or MINTUR) is responsible for the protection of all foreign tourists, the Ministry of Transportation (Ministerio de Transporte, or MITRANS) is responsible for mobilizing vehicles to assist with evacuations, and the Ministry of Public Health (Ministerio de Salud Pública, or MINSAP) is responsible for protecting the health of the population, such as ensuring sufficient equipment is available and testing water quality. The Council of Ministers, compiled of the ministers from each government ministry, works collaboratively to determine the most impactful means of disaster management. For decades, this has remained an important and distinctive characteristic of Cuba's approach.

Solidarity: A powerful component of the success of Cuba's model has been the implementation and promotion of solidarity. While a somewhat nebulous term, it can be defined as a "relation forged through political struggle which seeks to challenge forms of oppression" (Featherstone 2012, 5) or "a question of basic morality" (Appiah 2006, 157). For Cuba, its distinct brand of solidarity is rooted in its struggle against Spanish colonialism, and particularly the War of Independence (1895–1898), and well as U.S. neo-colonialism (1898–1958), as the country sought independence and its own national identity, or *Cubanidad*. In addition, the significant economic challenges faced by the population as a result of the U.S. embargo (since 1960, although not fully instituted until 1962), and later the collapse of the Soviet Union (1991) cultivated a deeply rooted and culturally ingrained understanding of the importance of solidarity.

Solidarity in post-1959 Cuba is, in essence, the view that human lives matter and as such it is the role of all Cubans to protect one another. Solidarity has been, and continues to be, tremendously important within the island's approach to protecting the population in times of crisis. This includes housing neighbors, cleaning debris, mobilizing, evacuating, and rescuing. While perhaps not a uniquely Cuban quality, it can be argued that the means by which solidarity is engrained in the national psyche, cultural memory, as well as how it manifests in times of crisis, solidary is indeed crucial to Cuba's approach to disaster preparedness and management.

Evolution: Perhaps the most distinguishing characteristic of Cuba's approach is its process of evolution. Much like the Revolution itself, Cuba's approach to disaster management and risk reduction has not been a staunch monolithic system. Rather, it has continued to evolve and adapt based on the population's needs, challenges faced, resources available, and changes in meteorological threats.

For example, since Flora made landfall in 1963, the disaster preparedness and management system has continued to change, develop, and improve. Much like the theme of prevention, this is evidenced throughout the system's individual components, as well as the system as a whole. Analysis was central to the ongoing advances in terms of, among others, the DCN's role, ministerial participation, support within the Constitution and legal framework, education, research and innovation, and strategies to combat climate change. The sweeping approach used today continues to undergo evaluation, as the Cuban government and population seek to perfect the system that protects lives and livelihoods. Evolution has been a consistent quality of the revolutionary project and is also represented clearly in its approach to mitigating natural hazards.

CONCLUSION

Since the 1960s, Cuba has undoubtedly developed an effective approach to dealing with natural disasters. Nonetheless, despite the success of Cuba's approach, challenges persist. Chief among challenges, for example, is the ongoing U.S. embargo. As Castellanos Abella and Wisner (2019) noted, "the economic blockade has placed severe burdens on the Cuban economy and the ability of the state to accomplish recovery in a timely way despite considerable mobilization of human and other resources." Another significant challenge is the cost associated with the damage, including structural damage. For example, between 2005 and 2016, hurricanes damaged 1,199,102 houses across the island and caused the total collapse of 147,710 houses. The total estimated cost of damage from hurricanes between 2005 and 2016 is $23.5 billion (Castellanos Abella and Wisner 2019). These figures are particularly significant when one considers the island, a middle-income, global south country, has a population of 11.3 million. In addition to the damage, another major problem is that recovery can be very time-consuming as a result of the annual hurricane cycle. In other words, it is challenging to completely rebuild an area that is often heavily impacted by storms, while simultaneously negotiating the need to work in the most devastated areas, leaving other damaged areas (although to a lesser extent) waiting for materials and recovery procedures. Nonetheless, overall, Cuba has created an impactful disaster preparedness and management system.

The key reason for the success of Cuba's approach has been its distinguishing qualities—making it a uniquely Cuban approach. This includes its characteristics that are distinctive from other national strategies from across the globe. These include Prevention; a Comprehensive National Civil Defense system; Education; Support of, and within, the legal system; Climate change recognition and adaptation; Involvement and participation of all Ministries; Solidarity; and Evolution. While the system as a whole is formulated by several uniquely Cuban characteristics, it has been extremely effective. This suggests that there are lessons which can be learned and strategies that can be adapted, to improve systems elsewhere. With the ongoing climate crisis and natural hazards increasing in intensity and frequency, creating adverse effects globally, the Cuban case offers innovative strategies. Indeed, while the model cannot be transplanted as such, there is undoubtedly a lot that can be gained from Cuba's experience.

NOTE

1. Stories of personal experiences during and after Hurricane Flora were collected throughout research for this study, largely from 2018–2020. In addition to formal interviews, research included conversations with friends and colleagues regarding their personal experiences and memories of Flora.

BIBLIOGRAPHY

Appiah, Kwanme Anthony. 2006. *Cosmopolitanism: Ethics in a World of Strangers*. New York: W. W. Norton & Company.

Carrasco Martín, Juana. 2012. "Datos preliminaries del impacto del huracán Sandy a su paso por Cuba." *Mesa Redonda*. October 26. Accessed September 20, 2019. http://mesaredonda.cubadebate.cu/mesa-redonda/2012/10/26/datos-preliminares-del-impacto-del-huracan-sandy-a-su-paso-por-cuba/.

Castellanos Abella, Enrique, and Benjamin Wisner. 2019. "Natural Hazards Governance in Cuba." *Oxford Research Encyclopedias*. Accessed January 30, 2019. http://oxfordre.com/naturalhazardscience/view/10.1093/acrefore/9780199389407.001.0001/acrefore-9780199389407-e-238.

Cubadebate. 2017. "Información del Consejo de Defensa Nacional." September 29. Accessed October 5, 2019. http://www.cubadebate.cu/noticias/2017/09/29/informacion-del-consejo-de-defensa-nacional/#.XLIKza2ZOt8.

Defensa Civil. 2017. *La Defensa Civil Cubana*. Havana: Casa Editorial Verde Olivo.

Featherstone, David. 2015. *Solidarity: Hidden Histories and Geographies of Internationalism*. London: Zed Books.

Figueredo Reinaldo, Oscar, and Dianet Doimeadios Guerra. 2017. "Tarea Vida: ¿Cómo enfrentará Cuba el cambio climático?" *Cubadebate*. 16 May. Accessed September 20, 2018. http://www.cubadebate.cu/especiales/2017/05/16/tarea-vida-como-enfrentara-cuba-el-cambio-climatico-video/#.XKdExq2ZOt8.

Galinsky, Seth. 2012. "Hurricane Sandy: Cuba Government Leads Effort to Meet Needs and Rebuild." *The Militant* 76 (42). November 19. Accessed September 20, 2019. https://www.themilitant.com/2012/7642/764202.html.

George Washington University. 2018. *Project Report: Ascertainment of the Estimated Excess Mortality From Hurricane María in Puerto Rico*. Accessed January 4, 2020. https://publichealth.gwu.edu/sites/default/files/downloads/projects/PRstudy

/Acertainment%20of%20the%20Estimated%20Excess%20Mortality%20from%20Hurricane%20Maria%20in%20Puerto%20Rico.pdf.

Granma. 2017. "Experts Warn of the Impact of Climate Change on Central Cuba." August 8. Accessed September 20, 2020. http://en.granma.cu/cuba/2017-08-08/experts-warn-of-the-impact-of-climate-change-on-central-cuba.

Guha-Sapir, Debarati, and Francesco Checchi. 2018. "Science and Politics of Disaster Death Tolls." *British Medical Journal* 362: K4005.

Houck, Oliver A. 2000. "Environmental Law in Cuba." *Journal of Land Use and Environmental Law* 16 (1): 1–81.

Kirk, Emily J. 2017. "Dealing with the Perfect Storm: Cuban Disaster Management." *Studies in Political Economy* 98 (1): 93–103.

Llanes Guerra, José. 2008. *Cuba: paradigma en la reducción de riesgo de desastres*. Accessed September 20, 2018. https://www.preventionweb.net/files/2558_CubaParadigmaenlareduccionderiesgodedesastres.pdf.

Mesa, Guillermo. 2008. "The Cuban Health Sector & Disaster Mitigation." *MEDICC Review* 10 (3): 5–8.

Thompson, Martha, and Izaskun Gaviria. 2004. *Cuba, Weathering the Storm: Lessons in Risk Reduction from Cuba*. Boston, MA: Oxfam America.

Pardo Guerra, Ramón, Luis A. Macareño Velíz, Albo Parra Salinas, Gloria Gely Martínez, Wilfred Cobas Dávila, Raúl R. Costa Gravalosa, Humberto de la Rosa, Noel Pérez Lozano, and Marbelis Rodríguez Azahares 2017. *Guía Metodológica Para la Organización del proceso de reducción de desastres*. Defensa Civil Cuba and UNPD. Accessed October 5, 2019. https://www.preventionweb.net/files/59362_guiametodologicaparaorganizacionrrd.pdf.

Puig González, Miguel Ángel, José Ernesto Betancourt Lavastido, and Rolando Álvarez Cadeño. 2010. *Fortalezas frente a huracanes (1959–2008)*. Havana: Editorial Científico-Técnica.

República de Cuba. 1976. *Constitución Política de 1976*. Accessed September 20, 2018. http://pdba.georgetown.edu/Constitutions/Cuba/cuba1976.html.

República de Cuba, Consejo de Defensa Nacional. 2005. *Directiva No. 1 Del Vicepresidente del Consejo de Defensa Nacional. Para la planificación y preparación del país par alas situaciones de desastres*. June 1. Accessed February 17, 2020. http://www.sld.cu/galerias/pdf/sitios/desastres/directiva_vp_cdn_sobre_desastres.ultima_version.pdf.

Santos Pérez, Santiago, Alfonso Núñez Leguen, Marbelis Rodríguez Azahare, Virginia Huergo Silveira, Milagros Santa Cruz Domínguez, Berta Lidia Castro, and Mayra Díaz Gaicía. 2018. *Buenas prácticas cubanas para la protección de niños y adolescents en situaciones de desastres*. Havana: Sello Editor de la Ecucación Cubana.

Smith, Rosi. 2018. "The 'Three Ps' (Perfective, Professionalization, and Pragmatism) and their Limitations for Understanding Cuban Education in the 1970s." In *Cuba's Forgotten Decade*, edited by Emily J. Kirk, Anna Clayfield and Isabel Story. Lanham, MD: Lexington Books.

UNISDR. 2015. "Sendai Framework for Disaster Risk Reduction." United Nations Office for Disaster Risk Reduction. Accessed September 20, 2019. https://www.unisdr.org/we/coordinate/sendai-framework.

Chapter 2

Disaster Management in Cuba

Formal, Semi-Formal, and Informal Procedures

Jessica Hirtle

Cuba's disaster management system, built over six decades, is extremely complex and multilayered. This chapter seeks to untangle these complexities by organizing the country's disaster preparedness and response measures into three distinct, but complementary, categories: formal, semi-formal, and informal. In this context, formal methods include government-led practices, semi-formal methods are defined as procedures run by community-based organizations, and informal methods involve interactions between individuals. All of these strategies require the involvement of a variety of actors, such as the Cuban government, National Civil Defense (Estado Mayor Nacional de la Defensa Civil, or EMNDC, or DCN), Civil Defense (Defensa Civil, or DC), Committees for the Defense of the Revolution (Comités de Defensa de la Revolución, or CDR), Federation of Cuban Women (Federación de Mujeres Cubanas, or FMC), and the Institute of Meteorology (Instituto de Meteorología, or INSMET), among others.

Despite repeated exposure to severe storms, such as hurricanes, Cuba's disaster management model continues to demonstrate the capacity to prepare for natural disasters and protect its population. Multi-level participation allows for the distribution of responsibilities, the coordination of different skill sets, and collective action, all of which create a streamlined disaster management system with expansive services. The efficacy of Cuba's disaster management system is grounded in these formal, semi-formal, and informal procedures, which have enabled Cubans to work collaboratively to mitigate disaster-related risks and minimize casualties.

FORMAL PROCEDURES

In Cuba, disaster preparedness and response procedures are firmly integrated within state policies and laws. Article 128 of Cuba's Constitution states that the President of the Republic has the responsibility to mobilize the population and announce a state of emergency. Furthermore, Article 137 asserts that the Council of Ministers is charged with the duty to protect the population and property in the event of a disaster (Constitution of the Republic of Cuba 2019).

It is the DCN, however, which acts as the driving force of Cuba's formal preparedness and response measures. Through governmental involvement, policy, and laws, Cuba has sanctified the authority and duty of the DCN to streamline and solidify the country's preparedness and response procedures (República de Cuba 1994). The National Defense Law No. 75, as well as Decree Law No. 170 of the System of Civil Defense Measures, regulate and authorize the responsibility of disaster preparedness and response within the DCN system. In the event of a state of emergency, such as a natural disaster, Law No. 75 declares that the National Defense Council can authorize the mobilization and evacuation of the population (Ibid.); it regulates that all resources are placed under the jurisdiction of the government to adhere to the needs of the DCN in the case of "exceptional situations" (Ibid.). Moreover, Decree Law No. 170 outlines that in the event of a natural disaster, other organizations, agencies, and institutions within Cuba communicate with the DCN and make their resources available to assist the population (Decreto Ley 170 1997). Decree Law No. 170 also stipulates that the DCN is required to work in coordination with other governing bodies, such as the Ministry of Agriculture (Ministerio de Agricultura, or MINAG), Ministry of Public Health (Ministerio de Salud Pública, or MINSAP), Ministry of Finance and Prices (Ministerio de Finanzas y Precios, or MFP), and the Ministry of Science, Technology and Environment (Ministerio de Ciencias, Technología y Medio Ambiente, or CITMA) to execute specific tasks in cases of natural disasters (Decreto Ley 170 1997).

The DCN is a government-run institution led by the president of the Council of Ministers (Llanes Guerra 2010). The main focus and foundation of the DCN, and the overall disaster management model, is the preservation of human lives, while the protection of material items is considered secondary (Sánchez Díaz 2018; Sánchez Mesa 2018). Enrique Castellanos Abella, director of geology in Cuba's Ministry of Energy and Mines (Ministerio de Energía y Minas, or MINEM), maintains this contention, noting that, "They do everything they can to avoid casualties. . . . Here, if somebody dies, it's big news. It's something outstanding. It's something really strange. The value of life here is very important for our culture" (Castellanos Abella 2018a).

In order to safeguard the population, disaster preparedness and risk reduction act as fundamental elements and are rooted within Cuba's overall disaster management system. With the support of the United Nations Development Programme (UNDP), the DCN established the use of Risk Reduction Management Centers (RRMCs) in 2005 (Llanes Guerra 2010). By 2015, RRMCs were located in eight provinces and 84 (of 169) municipalities throughout Cuba, working under the direction and supervision of the DCN to assemble information for risk-related studies, generate a culture of preparedness, reduce vulnerabilities, and promote risk reduction (Castellanos Abella and Wisner 2019; United Nations Development Programme 2015). Additionally, the RRMCs are tasked with making information available to a variety of institutions and organizations to foster knowledge exchange, the creation of tools, and improved communication (Llanes Guerra 2010).

Furthermore, every May—before the start of the hurricane season—the DCN organizes *Meteoro*, a two-day training exercise for all Cuban citizens (Mas Bermejo 2006). Since 1986, *Meteoro* has functioned as an organized and nationwide method of preparation for potential natural disasters and a comprehensive evaluation of the current disaster management system (Castellanos Abella and Wisner 2019). Its main objectives are to re-emphasize the steps that must be taken in the case of a natural disaster, to review changes that have been made to earlier protocols or practices, and to commence any hands-on preparations prior to the start of the hurricane season (Kirk 2017). During *Meteoro*, the entire population participates through their schools, workplaces, neighborhoods, and institutions (Almaguer 2018; Castellanos Abella 2018b). Citizens participate in exercises and simulations, which involve reacting to hypothetical disaster-related situations (Thompson and Gaviria 2004). Cuban officials and experts also evaluate the resources, risks, and deficiencies within communities. Preparations are also undertaken for the upcoming hurricane season, as steps are taken to mitigate damage, such as cutting trees and fortifying dams (Mas Bermejo 2006).

Although it is a national exercise, Angel Sánchez Riverón (2018), a specialist in Cuba's Civil Defense at the University of Havana, noted that each institution (including hospitals, government offices, and research centers) conduct their own analysis and exercises, which suits their specific needs and potential risks. There are three defense and reduction brigades at the University of Havana, for example, which prepare and respond to natural disasters and potential armed confrontations (Ibid.); in preparation for an impending storm, one of the brigades secures equipment, such as computers and tables. The first day of *Meteoro* is dedicated to the evaluation of the university as a whole. The director and deans discuss the actions completed that year and other potential preparations for the upcoming hurricane season. The second day includes cleaning the campus and reducing risks, such as clearing garbage and cutting down branches. The university also conducts practical exercises, including evacuations and rescues (Ibid.).

In the broader context, mass evacuations are a common and integral component of the overall formal preparedness and response procedures. For example, when Hurricane Ivan reached Cuba in 2004, 2 million people were effectively evacuated and zero fatalities were recorded (Thompson 2007). Similarly, Hurricane Ike in 2008 saw the evacuation of an estimated 2.7 million Cubans, while almost 350,000 people were evacuated in 2012 for Hurricane Sandy (Castellanos Abella and Wisner 2019; Sánchez Miranda and Choonara 2010).

Mass evacuations are used to relocate individuals who live in at-risk areas that are vulnerable to strong winds and flooding (Aguirre 2005), thereby preventing loss of life. When a hurricane is on course to reach Cuba, the DCN will order, organize, and lead the evacuations of the at-risk population (Thompson and Gaviria 2004). It collaborates with other organizations, including the FMC and CDR, to determine who needs to be evacuated. In this context, the DCN's primary responsibility, however, is to oversee evacuations and transport people to safety (Kirk 2017). All means of transportation are used to evacuate the vulnerable population, including buses, trucks, cars and, in high-risk situations, boats, and helicopters (Defensa Civil 2017). The evacuated population is then sent to *albergues* (shelters), which are housed in state-run institutions, such as schools or government buildings (Thompson and Gaviria 2004). Constructed and arranged by the DCN, the shelters are equipped with essential resources, including food, water, medicine, and blankets (Kirk 2017). Shelters are assessed by epidemiologists to ensure safe and sanitary living conditions and are staffed with medical personnel, police, Red Cross representatives, psychologists and, at times, even clowns to entertain children (Thompson and Gaviria 2004).

Communication is another key element of the Cuban disaster management model. INSMET informs the population of possible natural disasters through the transmission of weather-related information to the DCN and government, which is then conveyed to the local population. INSMET focuses on providing early warnings, which give the Cuban population 48–120 hours of notice of impending dangerous weather phenomena. It is essential for INSMET to cooperate and communicate with the DCN to allow "greater time for the preparation and execution of protective measures." These early warnings also allow citizens to "become aware of the danger" and therefore have the capacity to take precautionary and preventive measures (Teresita Llanes 2018).

The state-run Cuban media also plays an essential role in relaying information from INSMET and the government to the general population. Over the course of a given year, the media uses television programs and broadcasts to spread information on disaster preparedness, risk reduction, and response to Cuban citizens (Thompson and Gaviria 2004). In the event of a disaster, however, the media follows a well-organized course of action in transmitting information to the public (Castellanos Abella and Wisner 2019). The government utilizes Cuba's state-run media to communicate information, updates, warnings, decisions, and directives to the Cuban population (Kirk 2017). Weather updates are shared through a variety of media, including radio, television, and telephone (Meteorological Center of Sancti Spíritus 2017).

To protect the population and economy in times of disaster, and as noted in other chapters in this volume, in 1997 Cuba implemented a thorough four-phased plan of action: information, alert, alarm, and recovery (Decreto Ley 170 1997). In the event of an incoming hurricane, the DCN declares each phase and informs the media (Thompson and Gaviria 2004). The first phase, which is known as "Information," starts seventy-two hours before a hurricane is expected to reach Cuba. During this stage, information is communicated from the government to the media and then disseminated to the general population through television channels and the radio (Puig González et al. 2010). The media works under the guidance and instruction of the DCN to provide citizens with updates on the incoming hurricane, the current phase of action, the weather's impact in all regions of the country, the measures to be taken, evacuation plans, and any other relevant news (Thompson and Gaviria 2004).

The second phase, "Alert," begins forty-eight hours before the arrival of a hurricane. In this phase, the at-risk population is mobilized and evacuated. Schools are closed and students sent home, animals are moved to safety, and buildings are secured. If members of the population need to evacuate, the DCN, CDR, FMC and doctors work collaboratively to ensure that the at-risk individuals are protected and moved to designated shelters (Kirk 2017).

The third phase, "Alarm," is activated when the hurricane makes landfall and continues until the hurricane has passed over Cuba. During this time, all levels of the DCN remain in constant contact with one another via radio. Members of the DCN remain at their designated posts and await further directions or measures to be taken. Furthermore, the media provides storm updates to the general population, such as its location, strength, and projected path, through state-run news and radio. During this time, citizens are urged to find shelter and remain protected from the storm (Puig González et al. 2010).

After the hurricane, the Cuban population enters the fourth phase: "Recovery." In this stage, the DCN works to purify and determine the availability of potable water, assess structural damages, and inspect buildings and homes to ensure they are secure and inhabitable. The government then determines and announces when it is safe for the population to vacate their designated shelters and return to their homes. Cleaning as well as rebuilding and recovery strategies are also organized and implemented by the government and DCN, among others (Kirk 2017). Crews are mobilized to clean debris and services are restored. The extent of structural damages to buildings and homes are noted in a census to formalize and prioritize repair and recovery plans. A separate census also assesses and quantifies economic and agriculture-related damages, such as the loss of infrastructure, crops, or animals (Defensa Civil 2017).

In addition, MINSAP and, in particular, family doctors, work in collaboration with the DCN and other organizations, such as the Cuban Red Cross, to assist in Cuba's preparedness and response measures in order to safeguard the health and well-being of the Cuban population. It should be noted that the (state-employed) family doctors are crucially important in times of disaster. They are located in all Cuban communities and act as the "frontline" for health care, maintaining records of patients who are sick or have specific health-related needs (Defensa Civil 2017). Doctors share this information with members of the CDR to update risk-mapping and emergency plans. During a natural disaster, family doctors are stationed in specific areas and are charged with the following responsibilities: examining patients, helping with evacuations, and organizing hospital supplies. If a doctor is assigned to a shelter, it is their responsibility to inspect the shelter's conditions and replenish medical resources (Thompson and Gaviria 2004).

In assessing the efficacy of disaster preparedness, the formal role too of the National Education System cannot be overlooked. Primary to post-secondary students learn about the CDN and how to safeguard themselves and their communities in the event of a natural disaster through activities, such as drills and school exercises (EMNDC 2010; Castellanos Abella and Wisner 2019). In addition, topics related to the environment are also integrated within Cuba's national curriculum. For example, in grade three, students learn about nature through the "The World We Live In" program, in which they learn about potential hazards and how to protect natural entities, such as air, water, and soil (MINED 2019). While this program does not specifically cover disaster management, there is a plan in place to further incorporate the topics of danger, vulnerability, and risk into school curricula (Muñoz Campos 2018).

Formal procedures have also been put in place in response to the already visible and tangible impacts of climate change. Since the last century, the annual temperature in Cuba has risen 0.9° and sea levels have increased by 6.77 centimeters. It is projected that sea levels will rise a further 27 centimeters by 2050 and 85 centimeters by 2100 (CITMA 2017). Furthermore, Cuba was impacted by an "unprecedented" nine major hurricanes between 2001 and 2017 (CITMA 2017, 3). Rising sea levels and hurricane-related weather cause increased coastal flooding and erosion, which will lead to further loss of land and infrastructure (CITMA 2017); it is estimated that 5 percent of Cuba's coast will be lost by 2100 (Castellanos Abella and Wisner 2019).

In response to these projections, the Cuban government has implemented several national and local strategies to mitigate the impacts of climate change and raise public awareness of environmental sustainability. For example, *Tarea Vida* (Project Life), a national initiative to combat climate change, was introduced in 2017 (Rodríguez Milán and González Del Toro 2018). Headed by CITMA, the project outlines "actions to be taken" (Rodríguez Milán and González Del Toro 2018) in the short term (2020), medium term (2030), long term (2050), and very long term (2100) (CITMA 2017). Another formal initiative was developed with the view to ensure the safety of its population by building new houses further away from coastal and flood areas (EMNDC 2010).

Changes in construction techniques have also been introduced. For example, houses are regularly outfitted with metal and gabled roofs, which help resist strong winds, while concrete is reinforced with polystyrene to increase structural durability (EMNDC 2010). In addition, the government launched subsidies and low-interest loans, as well as grant schemes ranging from $200 to $3,000, to increase the accessibility of the necessary materials to fix and rebuild homes that have been damaged as a result of natural disasters (Rainsford 2012).

An additional and under-researched component of Cuba's formal strategy is knowledge sharing and cooperation, specifically regarding meteorology. INSMET, for example, collaborates with foreign nations, like the Dominican Republic, Haiti, and Equatorial Guinea. In these collaborations, INSMET shares its weather forecasting strategies, which include an early alert system, procedural manuals, and automated weather stations (Barbosa León 2016; Teresita Llanes 2018). The primary aim of this knowledge-sharing includes research and collaboration. As Miriam Teresita Llanes, chief of the National Weather Forecast Center in INSMET, noted, the institute receives members from foreign countries and sends its own representatives to conduct research on alternative strategies and methods (Teresita Llanes 2018). As she stated, "It's not about imposing our own opinion, it's not about imposing our own views. It's adapting what we do to their own specific countries. . . . We learn from the experiences of these countries" (Ibid.). Indeed, the strictly formal components of the island's disaster management strategy have been, and remain, significant.

SEMI-FORMAL PROCEDURES

Cuba's disaster preparedness procedures also involve the distinctive category of semi-formal methods. Although these are not formally managed by government institutions or entities, they nonetheless serve as a distinguishing and important part of Cuba's disaster management approach. Characterized by incorporating a mix of both governmental and local participation, the semi-formal component of the island's overall disaster management strategy has served as an additional layer of preparedness and protection.

As previously discussed, governmental institutions collaborate with mass organizations, such as the CDR and the FMC, to safeguard the population. Overall, the vast majority of Cubans are associated with the CDR and/or the FMC (Kirk 2017; Llanes Guerra 2010); 4 million women over the age of 14, for example, are FMC members (*Telesur* 2017). The CDR acts as an integral component of the Cuban model and works in collaboration with government entities to execute disaster preparedness and response procedures and disseminate information to the local population. The general duties of each CDR include recording population numbers within their designated areas. In relation to disaster preparedness and response, they utilize risk-mapping to determine the vulnerabilities of a given community (Thompson 2007). The CDRs keep detailed records of all neighborhood households, shelters, and hurricane-related vulnerabilities of individual homes (Thompson and Gaviria 2004). Additionally, each CDR also creates and updates a census to document which citizens will need help in the event of an evacuation and who will be able to provide assistance to those individuals (Thompson 2007). The FMC also assists in risk-mapping vulnerabilities by monitoring the details of all females living within a specific area (Thompson and Gaviria 2004). The collected information is then sent to, and used by, other disaster management groups, such as DCN leaders.

In addition to risk-mapping, CDRs also contribute to other preventative methods, such as helping to cut down trees that could damage telephone and power lines. Members also communicate measures to be taken in the event of a hurricane, such as a need to boil or save water. Moreover, as a result of their knowledge of specific areas, the CDRs work with authorities, particularly the DCN, to encourage and help evacuate local communities (Almaguer 2018). In other words, CDRs are used as a tool to transmit information to support the government and population in times of disasters.

It is also important to mention the culture of hurricanes, resilience, and safety that has been developed under the Revolution since 1959. Through the DCN, INSMET, the media, Cuba's education system and the Cuban Red Cross, citizens are educated on disaster prevention and response, "all of which form the 'Culture of Safety'" (Thompson 2007, 9). For example, José Rubiera, a noted meteorologist and director of the Cuban National Forecasting Center at INSMET, has played a key role in encouraging and continuing this culture of safety and prevention. Dubbed "chief weather forecaster," Rubiera is a valued and trusted source of information (Reed 2018). He hosted two television shows, "Global Weather" and "Weather in the Caribbean" and, in his nightly broadcasts, provided weather updates (Reed 2018). In times of disaster, such as hurricanes, Cubans regularly looked to Rubiera's special broadcasts for forecast updates. As Gail Reed explains (quoted in Hackel 2017), "Rubiera has been with Cubans over the decades, during every storm, every hurricane, every threat meteorologically to the island . . . they trust him because he teaches them."

As a result of the combined efforts of Rubiera and INSMET, among others, Cuban citizens are well-informed, aware, and knowledgeable on the topics of weather, prevention, and safety. It is clear that, through the development of Cuba's formal disaster preparedness and response procedures, it has been possible to build additional culturally focused methods, which continue to be shared and spread among the entire population. As a result, in the event of a natural disaster, Cubans know what to do and when to do it. Over the decades, citizens have equipped themselves with the tools to take precautions and prepare for natural disasters. While the formal structures noted earlier are more tangible and measurable, it is, nonetheless, important to identify and note the importance of the semi-formal strategies. While perhaps distinctly Cuban, they represent a vital competent to the island's success in disaster management, as formal, semi-formal, and informal engage collaboratively.

INFORMAL PROCEDURES

In addition to Cuba's formal and semi-formal procedures, acts of solidarity are common among the local population and can be explained as informal procedures. The island's particular brand of solidarity plays an instrumental role in the informal foundation and structuring of disaster management. Both the value and action of solidarity are deeply integrated within the country's relief strategies. As August Nimtz (2009, 133) stated, "Cuba's most essential feature is the presence and reproduction of human solidarity."

To explain, in response to the success of the rebellion, and subsequently the beginning of the Revolution, solidary values and behaviors were shared and spread among the population. This has allowed the concept of solidarity to become a value and, in turn, an action. Based on a complex mix of José Martí's ideology, Spanish colonialism, U.S. neo-colonialism, and Marxist-Leninist influences, Cuban solidarity can be characterized as having "genuine concern for each other's welfare" (Cureton 2012, 696). Unified by their shared characteristics and values, the majority of Cuban society cooperates together with the view to achieve and support shared goals (Cureton 2012; Ferdman and Kohn 2018).

In effect, Cubans are heavily involved in social networks related to the workplace and the communities in which they live. These social networks are highly integrated and interdependent, which creates and promotes the idea that all individuals are connected and inter-reliant. As Will Grant (2017) writes, there exists in Cuba a culture of "neighborly solidarity—ordinary people helping each other." Consequently, in times of disaster, there is a high degree of communication among local populations. Furthermore, it is evident that Cubans work collectively, often sharing supplies and finding refuge in each other's houses; it is typical for Cubans to "self evacuate" and stay in the homes of family members, friends, neighbors, or even strangers (Friedman-Rudovsky 2013). When Hurricane Ike passed through Cuba in 2008, for example, while an estimated 2.7 million people were evacuated, only 500,000 of them required shelter in formal government-led evacuation centers. The remainder of those affected stayed with relatives or neighbors (Castellanos Abella and Wisner 2019; Sánchez Miranda and Choonara 2010).

The value and action of solidarity promotes a clear desire and readiness to help, participate, and contribute, whether it be sharing resources with neighbors, the DCN, CDR, and FMC helping citizens evacuate, or medical professionals working in shelters. As one colleague noted, "Cubans do not have much, but they give what little they have" (Sánchez Mesa 2018). Inarguably, the informal processes, untouched and uninfluenced (formally) by the state, are rooted in solidarity and have had a significant impact on the development and implementation of disaster management.

THE SIGNIFICANCE OF CUBA'S FORMAL, SEMI-FORMAL, AND INFORMAL PROCEDURES

Cuba's formal, semi-formal, and informal practices act as key examples of effective disaster-related practices, which include components of multi-level governance, mobilization of the population, and the creation of social capital. A contributing factor to the efficacy of the country's disaster management strategies is the structuring and organization of the overall system. Disaster policy and close government involvement lead to an engaged and prepared system with defined and designated responsibilities (Sánchez Mesa 2018); these factors make it possible to formalize plans of action, assign duties, and collaborate with multiple sectors with differing expertise.

The leading role of the DCN, in collaboration with other organizations, such as the CDR, FMC, and INSMET, has resulted in a well-ordered and efficient system, with fixed responsibilities delegated evenly, varying from risk-mapping, risk reduction, evacuations, and communications. Through formal, semi-formal, and informal channels, virtually all members of society (the government, INSMET, health professionals, and the local population) play a role in preparing and responding to a given natural disaster. There is a general desire and readiness to help, participate, and contribute to the disaster management model, whether it be through sharing resources with neighbors, evacuation programs overseen by the DCN and mass organizations, or medical professionals working in shelters. Therefore, the utilization of social capital within Cuba allows for collective action concerning different skill sets and responsibilities, resulting in a more efficient structure with far-reaching capabilities.

Cuba's disaster management system also incorporates mass participation in disaster education. Through various channels, the entire population is informed and educated on hurricane-related risks, the importance of prevention, and how to react in the event of a disaster (UN Office for Disaster Reduction 2004). Such widespread awareness creates a culture of preparedness and efficiency, which, in turn, reduces panic, risk, and lives lost. The implementation of disaster education or prevention-focused programs allows the population to respond in a safe, effective, and efficient manner (Díaz González 2018).

At all levels of Cuba's disaster management system—from formalized institutional actors and programs, semi-formal organizations, and informal networks among the local population—Cuban citizens have the capacity to share, internalize, and understand disaster-related risks and preparation measures. Ultimately, a knowledgeable and involved population leads to adequate preparation, decreases vulnerabilities, and reduces the number of human casualties. Systematic, streamlined, and accessible communication procedures lie at the heart of such high levels of awareness and preparedness. Long-standing collaboration between the government, the DCN, INSMET, and the media allows for widespread and efficacious dissemination of essential information among the population in the event of a natural disaster, allowing citizens adequate time to take action to mitigate risks and/or to evacuate.

Within all components of Cuba's formal, semi-formal, and informal procedures—including RRMCs, *Meteoro* training, evacuations, and international response strategies—there lies a common thread: preservation of human life. In other words, people and their safety take priority over material resources. As Hidalgo López-Chávez (2018) noted, Cuba has "a very humanist system. What is important is to preserve your life. The most important thing is to preserve life, not things. That's what makes it wonderful." This prioritization of human life has seen Cuba minimize the number of disaster-related casualties to an impressive extent in comparison with other, similarly affected, nations. For example, in 2008, Hurricanes Ike and Gustav, both classed as Category 4 (CNN 2020; *The Guardian* 2008), struck Cuba, causing just seven deaths (Keyser and Smith 2009). Eight years later, as a result of Hurricane Matthew in 2016, over 1,000 people died in Haiti and 40 people were killed in the United States, but no lives were lost in Cuba (Kirk 2017). In 2017, several major hurricanes impacted the

Caribbean. In Puerto Rico, some 2,975 people died as a result of Hurricane Maria, another Category 4 storm (George Washington University 2018). In contrast, the Category 5 Hurricane Irma resulted in only ten deaths when it made landfall in Cuba in September 2017 (Althaus 2017).

Cuba's emphasis on safeguarding lives is also extended to other countries through its formalized international strategies. Even with minimal resources and financial capital, the country has stressed the importance of disaster diplomacy and works with other nations to provide disaster response services, with a focus on a shared sense of personhood and dignity. In nations around the world, Cuba assists relief efforts and works collaboratively to share knowledge and expertise, highlighting the importance of mutual exchange, community, and global ties. While it is the island's formal institutions that organize and respond to disaster-related situations overseas, its semi-formal and informal networks also uphold the same values and support international solidarity missions. As Castellanos Abella (2018a) explains, "the Cuban population has been educated in creating the willingness to help people from other countries."

CONCLUSION

In conclusion, this chapter has demonstrated that Cuba's formal, semi-formal, and informal procedures all act as integral and essential components of the island's overall disaster management model. Through policy, risk reduction, and mass participation, Cuba has formed a system in which all three categories work in conjunction to protect the population and respond effectively to natural disasters. The use of these three different, yet complementary, practices has created a multi-dimensional and systematic approach to disaster preparedness and response, with each component providing a specific service or contribution. Furthermore, these procedures all share a common objective and strive for the same outcome: to protect the population and reduce lives lost. Cuba's multilevel system utilizes the capabilities of all Cuban citizens and invests in social capital as a way to prepare and respond to natural disasters. The overall efficacy of the Cuban disaster management model, predicated on formal, semi-informal, and informal procedures, is demonstrated through its proven and recognized ability to mitigate and reduce risks and, most importantly, to minimize loss of life.

BIBLIOGRAPHY

Aguirre, Benigno E. 2005. "Cuba's Disaster Management Model: Should It Be Emulated?" *International Journal of Mass Emergencies and Disasters* 23 (3): 55–71.

Almaguer, Roberto. 2018. Interview with Jessica Hirtle. November 13. Havana.

Althaus, Dudley. 2017. "Hurricane Irma Blamed for 10 Deaths in Cuba." *The Wall Street Journal*. September 11. Accessed January 19, 2019. https://www.wsj.com/articles/hurricane-irma-blamed-for-10-deaths-in-cuba-1505148584.

Barbosa León, Nuria. 2016. "Cuba's Institute of Meteorology Highly Respected." *Granma*. February 9. Accessed January 19, 2019. http://en.granma.cu/cuba/2016-02-09/cubas-institute-of-meteorology-highly-respected.

Castellanos Abella, Enrique. 2018a. Interview with Jessica Hirtle. October 14. Havana.

Castellanos Abella, Enrique. 2018b. Interview with Jessica Hirtle. November 24. Havana.

Castellanos Abella, Enrique, and Benjamin Wisner. 2019. "Natural Hazards Governance in Cuba." *Oxford Research Encyclopedia of Natural Hazard Science*. Accessed February 15, 2020. https://www.researchgate.net/profile/Enrique_Castellanos2/publication/330635389_Natural_Hazards_Governance_in_Cuba/links/5c4b321fa6fdccd6b5c85750/Natural-Hazards-Governance-in-Cuba.pdf.

CITMA. 2017. *Enfrentamiento al Cambio Climático en la República de Cuba: Tarea Vida*. Ministerio de Ciencias, Technología y Medio Ambiente. Havana.

CNN. 2020. "Hurricane Statistics Fast Facts." June 2. Accessed September 15, 2020. https://www.cnn.com/2013/05/31/world/americas/hurricane-statistics-fast-facts/index.html.

Constitution of the Republic of Cuba. 2019. *Cuba's Constitution of 2019*. Accessed February 24, 2019. https://www.constituteproject.org/constitution/Cuba_2019.pdf?lang=en.

Cureton, Adam. 2012. "Solidarity and Social Moral Rules." *Ethical Theory and Moral Practice* 15 (5): 691–706.

Decreto Ley 170. 1997. *Del Sistema de Medidas de Defensa Civil*. May 8. Accessed February 15, 2020. https://www.ifrc.org/docs/IDRL/Cuba%20Defensa%20Civilcub82158.pdf.

Defensa Civil. 2017. *La Defensa Civil Cubana*. Havana: Casa Editorial Verde Olivo.

Díaz González, Elena. 2018. Interview with Jessica Hirtle. October 31. Havana.

EMNDC. 2010. *Civil Defense Cuba*. Havana: Estado Mayor Nacional de la Defensa Civil. Ferdman, Avigail, and Margaret Kohn. 2018. "The Theory and Politics of Solidarity and Public Goods." *Critical Review of International Social and Political Philosophy* 21 (5): 545–553.

Friedman-Rudovsky, Jean. 2013. "Hurricane Tips from Cuba." *New York Times*. July 29. Accessed April 17, 2020. https://www.nytimes.com/2013/07/30/science/hurricane-tips-from-cuba.html.

George Washington University. 2018. *Project Report: Ascertainment of the Estimated Excess Mortality from Hurricane María in Puerto Rico*. Accessed February 15, 2020. https://razoncienciaspr.files.wordpress.com/2018/08/informe-universidad-george-washington.pdf.

Grant, Will. 2017. "How Cuba and Puerto Rico Responded to Their Hurricanes." BBC News. September 23. Accessed Accessed April 17, 2020. https://www.bbc.com/news/world-us-canada-41371793.

Hackel, Joyce. 2017. "To Reassure Cubans, a Revered Meteorologist Returns to the Airwaves." *The World*. September 12. Accessed Accessed April 17, 2020. https://www.pri.org/stories/2017-09-12/reassure-cubans-revered-meteorologist-returns-airwaves.

Hidalgo López-Chávez, Vilma. 2018. Interview with Jessica Hirtle. November 9. Havana.

Keyser, Jonathan, and Wayne Smith. 2009. "Disaster Relief Management in Cuba: Why Cuba's Disaster Relief Model is Worth Careful Study." *International Policy Report*. Washington: Center for International Policy.

Kirk, Emily. 2017. "Alternatives–Dealing with the Perfect Storm: Cuban Disaster Management." *Studies in Political Economy* 98 (1): 93–103.

Llanes Guerra, José. 2010. *Cuba Risk Reduction Management Centres: Best Practices in Risk Reduction*. Colombia: United Nations Development Programme.

Mas Bermejo, Pedro. 2006. "Preparation and Response in Case of Natural Disasters: Cuban Programs and Experience." *Journal of Public Health Policy* 27 (1): 13–21.

Meteorological Center of Sancti Spíritus. 2017. "Fortalecimiento del Sistema de Alerta Temprana (FORSAT)." Forecast Brochure. Sancti Spíritus, Cuba.

MINED. 2019. "Ministerio de Educación: República de Cuba." Accessed Accessed April 17, 2020. https://www.mined.gob.cu/.

Muñoz Campos, Marta Rosa. 2018. Interview with Jessica Hirtle. October 19. Havana.

Nimtz, August. 2009. "Cuban Solidarity and Disaster Response." *Latin American Perspectives* 36 (2): 133–134.

Puig González, Miguel Ángel, José Ernesto Betancourt Lavastida, and Rolando Álvarez Cedeño. 2010. *Fortalezas frente a huracanes*. Havana: Editorial Científico.

Rainsford, S. 2012. "Cuba's Crumbling Buildings Mean Havana Housing Shortage." *BBC*. May 17. Accessed Accessed April 17, 2020. https://www.bbc.com/news/world-latin-america-17935769.

Reed, Gail. 2018. "Weathering US–Cuba Political Storms: José Rubiera PhD Cuba's Chief Weather Forecaster." *MEDICC Review* 20 (2).

República de Cuba. 1994. Ley No. 75 de la Defensa Nacional. December 21. Accessed February 15, 2020. https://www.cubahora.cu/uploads/documento/2017/08/10/ley-no-75-de-la-defensa-nacional.pdf.

Rodríguez Milán, Yisell, and Danae González Del Toro. 2018. "Climate Change Brings Transformations in Cuba." *Granma*. April 12. Accessed April 17, 2020. http://en.granma.cu/cuba/2018-04-12/climate-change-brings-transformations-in-cuba.

Sánchez Díaz, Anna. 2018. Interview with Jessica Hirtle. October 31. Havana.

Sánchez Mesa, Belkis. 2018. Interview with Jessica Hirtle. October 23. Havana.

Sánchez Miranda, Deybis, and Imti Choonara. 2010. "Hurricanes and Child Health: Lessons from Cuba." *Archives of Disease in Childhood* 96 (4): 328–329.

Sánchez Riverón, Angel. 2018. Interview with Jessica Hirtle. October 29. Havana.

Telesur. 2017. "Federation of Cuban Women Marks Over 5 Decades of Equality." August 23. Accessed May 19, 2020. https://www.telesurenglish.net/news/Federation-of-Cuban-Women-Marks-Over-5-Decades-of-Equality--20170823-0042.html.

Teresita Llanes, Miriam. 2018. Interview with Jessica Hirtle. November 21. Havana.

The Guardian. 2008. "Hurricane Gustav Hits Cuba." August 30. Accessed Accessed April 17, 2020. https://www.theguardian.com/world/2008/aug/30/naturaldisasters.caymanislands.

Thompson, Martha. 2007. *Enhancing Urban Safety and Security: Global Report on Human Settlements*. London: Earthscan

Thompson, Martha, and Izaskun Gaviria. 2004. *Cuba, Weathering the Storm: Lessons in Risk Reduction from Cuba*. Boston, MA: Oxfam America.

UN Office for Disaster Reduction. 2004. "Cuba: A Model in Hurricane Risk Management." September 14. Accessed August 20, 2019. https://www.un.org/press/en/2004/iha943.doc.htm.

United Nations Development Programme. 2015. "A Cuban Model for a Resilient Caribbean." February 15. Accessed October 21, 2020. https://www.undp.org/content/undp/en/home/presscenter/articles/2015/02/25/a-cuban-model-for-a-resilient-caribbean.html.

Chapter 3

First and Last Bulwark against Natural Disasters

Cuba's Revolutionary Armed Forces

Hal P. Klepak

This chapter aims to describe, set in context, and analyze the place of the Cuban Revolutionary Armed Forces (Fuerzas Armadas Revolucionarias, or FAR) in the arrangements and approaches that the Cuban state has made for dealing with natural disasters on the island. It will do this by first giving a description of the nature of the challenges faced by Cuba in dealing with natural disasters, their types, frequency, and severity, then by giving a brief look at the Cuban military itself, and finally by assessing the position the FAR has in the state's response to the demands placed on them by the phenomena.

THE NATURAL DISASTER CHALLENGE IN CUBA

Cuba is a long and narrow island, some 109,984 square kilometers in size, and stretching over 1,200 kilometers from its western extension at Cape San Antonio on the Yucatán Channel and the Gulf of Mexico to its easternmost spot at Maisi on the Windward Passage separating it from the neighboring island of Hispaniola. Its northern coast bathes in the waters of the North Atlantic Ocean while on the South coast it is the Caribbean Sea which it faces (Núñez Jiménez 1982). In fact the country is an archipelago made up of the main island, a major second island of some 2,200 square kilometers, and a great many small keys. It is divided from the United States to the northwest by the Straits of Florida, to the northeast from the Turks and Caicos Islands, a British territory, to the southeast from Jamaica and to the southwest from the British territory of the Cayman Islands. From this geographic situation, the island receives its warm climate but equally its exposure to several kinds of natural disasters. And while it is the largest island and country in the Caribbean Basin, it is particularly susceptible to some serious climatic conditions leading to natural disasters as a result also of its being a mere 191 kilometers wide at its widest point and some 31 kilometers at its narrowest point.

The island is thus southerly but not tropical. Its exposure to natural disasters, however, is not merely a result of the purely geographic factors to which reference has just been made. It is also as a result of socioeconomic factors related to its colonial and, what is called in Cuba, its "pseudo-colonial past."[1] A list of threats of a natural kind to the island is a long and diverse one, and when disasters not strictly of a natural kind are added, it is lengthy indeed, including: hurricanes; tropical storms; extra-tropical lows; severe localized storms; intense rains caused by other factors; coastal flooding; serious droughts; earthquakes; rupture of hydraulic works; both urban and rural major fires; epidemics; serious agricultural plagues and diseases; hydrocarbon or other chemical substance spills; toxic gas leaks; radiological accidents; major explosions; and major catastrophes in transport (Álvarez Calvo 2015).[2]

Hurricanes of course have been the most discussed of these phenomena, but the nation has not entirely escaped any of those on this list in recent years. Christopher Columbus was the first European to visit the island, and he noted already on his third visit the risk presented by hurricanes, "these strong storms," originating in the Eastern Atlantic, that were so prevalent between May and November of each year. He wrote that "only for the service of God or by extension of the monarchy, would I again expose myself to such peril" (Ibid., 38). From the beginnings of Spanish settlement in the 1510s until the end of formal colonial rule in 1898, the island was seemingly hit over 200 times by the phenomena, since while we do not have figures for most of that period, we do, for example have them for 1891–1998 when some 143 struck Cuba. That is an average of just under 1.3 a year. Some historians even suggest that the lack of attention to the preparation for, and recovery from, such events had a significant role in creating resentment among the colony's population about Spanish rule and contributing to the growth of the independence movement. Be that as it may, Cuba has known these ferocious storms for many centuries (Francisco Álvarez 1994).

As if this risk were not enough for the island, added to the threat from hurricanes are what are termed "extra-tropical storms," these having their origins in the Gulf of Mexico, and present in the Caribbean near Cuba in the months from December to April. While rarely of the intensity of hurricanes, they are often very strong indeed, bringing high winds and great amounts of rain to the country and causing on occasion significant destruction and loss of crops. Thus, for essentially the whole year, Cuba is exposed to the potential presence of powerful and worrying storms. When the effects of the phenomenon of the "Niño" are considered as well, storms are truly at the heart of the risks the island faces permanently. The Niño's tendency to produce intense rains, resulting in serious flooding, important entrances of the sea into coastal areas, and conversely its propensity to also bring periods of lengthy droughts, make it yet another consideration of a high order in Cuban planning, although not in as steady a way as hurricanes and extra-tropical storms.

The major study on natural disasters of recent years, already cited, does not hesitate in saying that in the end, the "impacts of strong winds are the major source of danger in our country with marked effects on the population, areas of agricultural and industrial production and ecosystems throughout the whole country" (Álvarez Calvo 2015, 27). While other phenomena are important to consider in the realm of natural disasters in Cuba, the hurricane (and other high-wind producing storms) is king and must receive the lion's share of the nation's attention.

These other disasters are nonetheless important and often frighteningly so. Droughts are usually considered the most damaging of these other threats where Cuban attempts to deal with natural disasters are concerned. These abnormal but too common lengthy periods of dry weather and lack of water are especially noteworthy on the island. Cuba has, for a small country, a significant number of differences often noted between the climatic conditions of one region and those of another, often between even neighboring jurisdictions (Suárez Rivas 2014). There is frequently drought in one area and abundant rainfall a few kilometers away. Starting with a mere lack of rainfall, the problem soon can have higher temperatures prevailing, the presence of higher winds than normal, fewer clouds forming, and high solar radiation. Thus, evaporation increases.

The phenomenon is of course global with desertification a major problem in much of the world in recent times. Latin America has been no stranger to the problem with the region considered to be affected to the extent of one-quarter of its area (Álvarez Calvo 2015; Suárez Rivas 2014). The Caribbean is no exception to this global and regional context. In Cuba, it is a problem generally experienced at the end of winter and beginning of spring, in the months of March, April, and May when temperatures rise but the summer rains do not arrive as expected. It is usual for even larger areas of land to be affected than with flooding. The impact on agriculture and the economy in general can be enormous and if sustained livestock are affected, what in Cuba is called an "año de vacas flacas" (literally, a "year of thin cows"), lands can be abandoned, and many negative social impacts abound.

Related or not to droughts are fires, both urban and rural. Fires are often not natural disasters, in that a large percentage of them are man-made, but their impact becomes similar. Cuban authorities give the figure of 20 percent of fires being the result of nonhuman causes, such as electric storms, extreme meteorological conditions, high temperatures, low relative humidity, lack of rain, and/or the presence of drought conditions (Álvarez Calvo 2015).

The other phenomena listed above are usually the results of those just discussed. Flooding, the entrance of the sea onto drylands, and the others tend to be products of these main threats and above all, for Cuba, of hurricanes and other fierce storms. The exception is earthquakes. In recent centuries, the presence of earthquakes in Cuba has received relatively little attention. Those there have tended to be in the very Far-East of the island, and of a low scale with little or no damage to infrastructure and no loss of life. Little wonder then, in a country with so many constant difficulties in this field, that this type of natural disaster has been considered secondary. Having said this, the catastrophic Haitian earthquake of 2010 brought tremors to Eastern Cuba and since then there have been several others felt in that part of the island and even farther West. Increasing attention is now being leveled at them.

THE ARMED FORCES OF CUBA

With the challenge briefly laid out above, one can now turn to one of the principal instruments Cuba has at its disposal to answer the needs presented by natural disasters. These are the FAR, grouped under their ministry which is termed the Ministry of the Revolutionary Armed Forces (Ministerio de las Fuerzas Armadas Revolucionarias, or MINFAR) and not under a defense ministry as in most countries. The three armed services, Army, Navy, and Air Force, make up the FAR. While separate, the point should be made immediately that their history and roles have made them eschew most of the infighting known between rival services in many countries, and perhaps especially in Latin America. The three services work very closely together in almost all roles and natural disasters is no exception.

The forces are relatively new in make-up, being formed from the nucleus of the rebel army (Ejército Rebelde) which Fidel Castro led to victory in January 1959 over the old armed services of the Fulgencio Batista dictatorship. Those armed forces were disbanded completely after the Castro victory and only a relatively small number of personnel, whose records on torture and similar crimes were clean, were retained in the new forces.

The new military was to be truly revolutionary and serve the cause of the Revolution which Fidel and his followers quickly set about implementing in the months and years after their victory. It was not a-political but rather avowedly political, however only in the sense of being the servants of a revolutionary project. Indeed, it was dissimilar to the usual Latin American forces, steeped in the practice of involvement in politics, making and unmaking of governments, sabre rattling, *pronunciamientos* and *coups d'état*, and generally what are called in Latin America *deliberantes* (deliberative bodies). Instead, they have followed the leadership of their *comandante en jefe* (commander-in-chief) in peace and war with admirable discipline.

This has placed in the hands of the government a flexible, efficient, loyal, hard-working body of men and women, regulars and reservists, who are not surprised to be asked to do a vast range of tasks which many armed forces, especially those with more Prussian leanings, would consider non-traditional or even unacceptably below their dignity. As in most countries, the FAR are a unique part of the state apparatus bringing to the highest authorities in the land a wide range of capabilities and qualities not found elsewhere. They are:

- Loyal
- Accustomed to receiving orders and executing them without discussion
- Disciplined
- Hierarchically organized
- Constantly brought up to date through courses
- Accustomed to teamwork
- Accustomed to difficult conditions (weather, altitudes, etc.)
- Mobile
- Highly developed capacity to communicate
- Transportation capacity
- Present throughout the national territory
- Know the national territory
- Numerous
- Generally physically fit
- Quickly and widely deployable
- Contain within the organization-wide medical capacities
- Contain dental capacities within the force
- Contain postal capacities within the force
- Hold reserves of materiel useful in emergencies or conflict

It is not surprising that the military play such an important role in so many societies when one considers all these advantages and when one compares them with other institutions of the state, especially those of a developing nation.

In Cuba, the only institution that can be considered even in the same league in terms of capacities is the Communist Party of Cuba (Partido Comunista de Cuba, or PCC) itself. But even the most proud member of that organization would not claim that its list of advantages, at least in times of emergency, is in any way as long as that of the military. Fortunately, as will be seen, in Cuba the cooperation of these two central institutions is the key to successful natural disaster preparation, mitigation, and relief.

Since 1959, the FAR has been asked to undertake any number of missions outside the usually conceived roles assigned to traditional armed forces. When the capital fell to the rebel forces, and the new government daily proved that it was indeed going to implement a revolutionary program of change in Cuba, elements of the Cuban society of the day, fearing for their privileges, left the island and established themselves in other countries, mainly the United States. They expected that, as in the past, the United States would not long tolerate a deeply reformist government so close to its borders and that they would be restored to their running of the island in short order (del Pilar Díaz Castañón 2001).

Thus, the country was soon left without its former army, its aristocracy, much of its middle class, and its bureaucracy. Fidel could count on only one body of people to fill the huge gaps in state and society and that was the FAR. Thus, from the time of victory, the military became engaged in running much of the administration, including, crucially, the land reform program. If efficiency was not always impressive, loyalty to Fidel and to the revolutionary program certainly were. This tradition has continued to this day with the FAR being involved in fisheries, farming, industrial management, construction projects, and more recently, with the Special Period of 1990 shattering the country, tourism, mining, management of all manner of firms, tobacco, and a host of other fields of endeavor (Klepak 2011).[3]

THE FAR AND NATURAL DISASTERS

One of the first of such new roles was dealing with natural disasters striking the island. As in most countries, even the old armed forces of what is called in Cuba *la tiranía* (the Batista regime 1952–1958) were often involved in natural disaster relief. When the disaster came, lacking any other means to deal with its ravages, the armed forces were called out to assist with resettlement, temporary housing, feeding the displaced, and so on. But this effort was unplanned, unorganized, and haphazard to say the least, rather, unfortunately, like it still is in many countries in the Caribbean Basin (García Moreno 2011). The results tended to be high casualties, much destruction and slow recovery but precious little in the way of lessons learned for the next occasion.

Soon after the revolutionary victory, Cuba was struck by a series of hurricanes, some of them very ferocious indeed. The new government was determined to show its concern for communities hard hit by the storms, places usually forgotten by both colonial and post-colonial governments. Fidel and other leaders toured the devastated areas and spent real time in them, living under canvas and directing recovery efforts. The beginnings of the remarkable Cuban system of managing natural disasters were being laid (Puig González et al. 2010).

On July 31, 1962, Popular Defense (*Defensa Popular*), the first organ responsible for managing natural disasters, was founded on Fidel's order and with his brother Raúl, the minister of the Revolutionary Armed Forces since September 1959, the principal person responsible for its establishment and functioning (Cobas Dávila 2014).[4] Despite the interest of both Fidel and Raúl in the project, preparations for the defense of the island in the face of U.S. invasion or domestic insurrection meant that for the first years Popular Defense got rather less attention than was to be the case later on. Both the Bay of Pigs invasion of April 1961 and the October Missile Crisis of 1962 stole the headlines and the bulk of the resources the new government could give to national emergencies (Díez Acosta 2008; Etcheverry Vázquez and Gutiérrez Oceguerra 2008; Herrera Medina 2006; Kornbluh 1998).

Hurricane Flora, however, hitting the island in October 1963, changed all that. In the then province of Oriente, the devastation was horrific. Some 1,200 people died, hundreds more were injured, and huge property loss was suffered. Fidel was more active than ever visiting the stricken zones and his personal interest in building something new and more efficient became obvious. The FAR had done all it could to assist, and there had already been an exceptional improvement in assistance given, but now a real effort to change the whole nature of natural disaster management began. Never again was there to be anything like such loss of life from a hurricane in Cuba, at least to date. Disaster risk management was to be a priority of the new government. "Risk management" was defined in an original fashion and given meaning. For Cuba and the FAR, "the management of risk is a social process whose objective is knowledge, reduction and permanent control of the risk of disaster in society" (Defensa Civil 2017). Law 75 and Law Decree 170 soon set the FAR's role in disaster risk management, and indeed that of the subordinate Ministry of the Interior (Ministerio del Interior, or MININT), into a legal context. And this formal legal basis for FAR activity was in practice to be even greater than that assigned to in early documents (Álvarez Calvo 2015).

What other factors had meanwhile changed, however, were that the very defense arrangements the country was putting into place to face internal and external armed opposition was giving it the wherewithal to overhaul completely the natural disaster management system. Facing the threat and then the reality of U.S. or U.S.-sponsored invasion, as well as growing opposition domestically to reforms, especially the hotly resisted agrarian reform program, the Cuban government had little option but to mobilize public support for military preparation and action.

Cuba was not in a position to train and particularly to arm the hundreds of thousands of volunteers who presented themselves for the expanded regular and reserve forces being raised at the time. Fidel had hoped that his government would be accepted, if not liked, by the ruling elites and most especially by the United States. He had thought a regular army of perhaps 16,000 men would be enough in a country with stability and improving social justice. This was of course not to be. The United States soon plotted his downfall and the Catholic Church, the oligarchy, the increasingly large and powerful exile community, and what was left of the upper-middle-class and recalcitrant Batista security forces: all formed yet again in Latin American history the "Unholy Alliance" of conservative elements so well known in the past as blocking even the most moderate of reform efforts (del Pilar Díaz Castañón 2001; Morley 2002; Paterson 1994).

With the United States refusing to sell arms itself to the new government, and with its pressure on allies soon ensuring that they would not be doing so either, Cuba slowly but surely was pushed into arms and training deals with the Warsaw Pact, first with Czechoslovakia but then with the Soviet Union itself. A plethora of new reserve army organizations was being spawned, most under government initiatives but some springing up from the public itself (Rueda Jomarrón 2009). The government knew it could not defeat a serious invasion but it developed a strategy of deterring such an attack by making its cost far outweigh any advantages coming from it to the aggressor. This could only be achieved by a "nation in arms," building on the French Revolutionary model, and to an even greater degree on the more recent Vietnamese model (Nguyen Giap 2008; Pavlov 1993).

Cuba moved by 1965 to have full obligatory military service for its male population and a number of schemes whereby women would also do both regular and reserve service. Women had indeed been part of the fighting elements of the rebel army in the East before the victory of 1959 and the military support of women to the cause of national defense would be a major part of the overall effort. This was especially the case after victory through the efforts of Vilma Espín, Raúl's equally revolutionary wife (Waters 2003). This involvement of women in the defense posture of the nation was to pay off in large measure when natural disasters struck later on. From the beginning of the National Revolutionary Militias (Milicias Nacionales Revolucionarias, or MNR) and other reserve organizations in 1959, women were to have a full role in national mobilization in emergencies. And the gender perspective in national efforts in this natural disaster management has been increasingly noticeable as a result of its obvious advantages in this as in so many other fields of endeavor (Millán Álvarez 2014).

The island was divided into three great military zones: the Western Army, the Central Army, and the Eastern Army and into national defense "sectors."

These last were assigned to lower administrative divisions and organized along provincial and municipality lines. This structure worked for all matters of defense from applying conscription regulations to mobilization in case of war. This would be the bedrock on which the future Civil Defense (Defensa Civil, or DC) would be constructed. Defense Councils (Consejos de Defensa) would be set up at every level of government and administration: national, provincial, municipality, and even neighborhood/community. All state, military, and popular mobilization in the face of crisis would be their responsibility, and their most frequent activity would be in the field of natural disaster response.

Here is the living tissue of Cuba's response to this threat. Frequently exercised and trained, including everyone involved from the armed forces to the Red Cross, passing through MININT, the mass organizations, fire brigades, civil government, and the other key agencies, these councils can be activated quickly and go into action speedily. Since they exist at all levels, they ensure a coordinated response to the challenges of alerting the population and going into action from the first phase of preparation for the oncoming challenge. Especially in early evacuation and rescue efforts, they are central elements, and their claims on the armed forces and their resources are an essential part of all response (Cobas Dávila 2014).

The FAR is responsible for the proper functioning of the system. Since their priority responsibility over the last six decades has been deterrence and national defense in the context of a full "nation in arms," and this has obliged them to organize a system of defense throughout the country and engaging a huge percentage of the population, their traditional defense structure took in territory, people, and resources at all levels. The fact that they were accustomed to non-traditional roles both in the years of the rebellion and those following victory meant that a further role in something as vital to the nation's wellbeing as natural disaster mitigation proved neither surprising nor daunting to them. While the country was poor, the emphasis on national defense and furthering the goals of the Revolution provided the ethos and wherewithal to set up a real system to deal with the challenges posed by oft-seen events.

In the heyday of the Revolution's defense structure, the armed forces could count on well over 200,000 regular personnel. Conscription brought into the FAR most young men and retained them in uniform for three years. Afterward they were sent to the reserves and were liable to call-up for many further years. But many who did not serve as regulars were still in mass organization (*organizaciones de masas*), which received some military training.

Strictly speaking, and again under the FAR, there were also the various active militia organizations, and then there was, since the early 1970s, the Youth Labor Army (Ejército Juvenil de Trabajo, or EJT), a roughly

100,000-man body engaged in farming and producing much of the island's food supply, but first undergoing basic (*previa*) military training and wearing uniform throughout their normal conscripted service (Castro 1973; Forteza 2003). Defense Brigades (Brigadas de Producción y Defensa, or BPDs) were also created, functioning as local factory or agricultural unit bodies volunteering to defend the Revolution against internal and external enemies. These organs were to be civilian in dress but would come under MINFAR in case of crisis. They have been the base on which many popular mobilizations of support for the Revolution have been organized. They are true mass organizations and their manpower is impressive in numbers and in utility to the state. Their utility was proven dramatically in the mobilization to face the *Maleconazo* of August 1994 when mass rioting occurred in Old Havana.

Linked to these are the much-criticized Committees for the Defense of the Revolution (Comités de Defensa de la Revolución, or CDR). Set up in September 1960, in response to anti-Castro terrorist bomb attacks, the CDRs are organized at block and neighborhood levels, as well as with higher headquarters, and their official role is to combat any signs of counter-revolutionary activities. Modeled closely on the Eastern European communist model, they aim to discover cases of corruption, anti-social or illegal activity, and the like but they also have come to help with elections, organizing blood banks, fighting common delinquency problems locally, getting conscripts off to do their national service and welcoming them back home after that service is done, organizing participation in mass mobilizations, arranging the neighborhood street parties for national days, and much else (Malmierca 2005). They are ubiquitous across the nation, as present in rural areas as in the cities and towns of the island.

All these mass organizations respond either directly to MINFAR or, through MININT, more indirectly to the chief defense and security headquarters that MINFAR represents. Thus they can be mobilized easily and speedily through the military communications links that keep national defense major functions on track on a routine basis. They are also part of the Central System of Vigilance (Sistema Único de Vigilancia), organized in the early 1990s to face the growing challenges of internal order caused by the hardships of the Special Period. They constitute at their best a simply massive availability of manpower, to some extent disciplined but certainly available, to assist the regular forces in dealing with crises and the most common of such crises in Cuban history have been natural disasters and especially hurricanes (Lezcano Pérez 2008). Thus Cuba, almost uniquely, has this enormous body of often experienced men and women ready to be deployed in each phase of preparation for, dealing with, and mitigating the effects of natural disasters.

AND HOW IS IT ALL DONE?

Cuban Civil Defense organizes its response to disasters in three phases, not especially different from the practice in most other countries. There is the Recognition Phase (*fase de reconocimiento*), which consists of collecting and analyzing all of the information available on the phenomenon being faced and on its potential consequences, all of this in the shortest time possible. The intent is to provide each level of decision-making with concrete summaries on which to base their decisions and thus lead to the most rapid and effective response possible.

Typical issues dealt with are the numbers of people and livestock likely to be affected by the disaster, their location and special conditions, the extent of the area likely to be affected and the degree of seriousness of that effect, the likelihood of epidemics or other medical situations potentially arising, and an inventory of likely needs once the disaster has struck. Assessments deal with issues such as the condition of roofs, agricultural infrastructure, electrical cables and installations, means of transport and the need for their use in evacuations, access to water, and even such things as the likelihood of problems of loose dogs after the effects have hit. The desired end is the most rapid collection of information possible and that it be as "detailed, accurate, reliable and timely as possible" (Álvarez Calvo 2015, 20–21).

The next phase, the Emergency Phase (*fase de emergencia*), deals with identifying the actual zone hit by the disaster, its circumstances, and the possibilities of putting into practice an adequate response with those resources available. The first priority will be to, if necessary, evacuate and temporarily house, but in any case feed the affected population, and provide it with drinking water. This applies especially to the most vulnerable sectors, not merely by age but, for example, those in most need of milk. In addition, other vital needs, such as medicines, shelter, movement, and transport must be addressed.

In the next phase, the Recovery Phase (*fase de restablecimiento*), there is a major effort to discover the extent of the damage done, to continue to address the most affected part of the population, to ensure public health, to plant crops that can be harvested in the least time possible, and to engage in related measures that can lead to as rapid and complete a recovery as possible.

The roles of the armed forces in these three phases are many. They can be called upon in each phase and are part of the coordinated response of the nation, province, municipality, and even neighborhood. The regular armed forces are to some extent the least affected of the many organizations responding to MINFAR. As mentioned, they are much smaller than in the pre-Special Period past, having been cut from well over 250,000 to their current strength of perhaps 50,000 to 60,000.[5] Their manpower is thus not as massive nor is their presence in the national territory as widespread as in the past. At the same time, their resources in terms of medical capacity, engineering, transport and logistics (especially evacuation and accommodation), communications, and other vital needs ensure them a clear role.

The DC, with its national headquarters in Havana, is commanded and staffed by uniformed regular FAR personnel. They have the principal role in long-term planning, especially contingency planning, national preparation, and liaison with other interested agencies in the periods between emergencies. At the provincial level, the same tends to apply although the presence of reserve personnel from the militias and other mass agencies is more visible. And this increase in the percentage of non-regulars grows as one descends the chain of command to more local centers.

This is not obvious to the casual observer. When warnings, instructions, appeals, calls for assistance and the like are issued locally or provincially, the persons speaking are usually in uniform and wearing military ranks. But this must be seen in the context of a vast reserve of civilians who are still members of one of those reserve organizations. In addition, the Party, which also has a major role in mobilization in times of disaster, although one of less importance than the military's, will often be present with a member who is also a reservist and who will attend such meetings in uniform. This is the result of the mass engagement of non-regulars in the myriad organizations which work on natural disasters. Older members are, or were, almost always members of the reserve forces as well as of the Party.

It is this organization, made up of regulars, reservists of many stripes, members of mass organizations, like the CDRs and the BPDs, and other volunteers that ensure Cuba's exceptional system actually works. And the FAR are at every level active. In the Recognition Phase, their alert system is a vital part of the national prediction and weather reporting arrangements. Likewise they are key to the quick and accurate study of the current and potential situations for which decision-makers must prepare and then take action. Because of the detailed work previously done in preparing any number of contingency plans, at all levels, many of them are already on the move organizing a wide variety of activities, such as the chopping down of tree branches that might later come down on electric wires during a storm, building dikes, shoring up endangered slopes, and all the vast range of things to be done as the danger approaches. Others are investigating reports of endangered or weakened buildings and other facilities, especially those which might house potentially dangerous chemicals or other materials, and attempting to reinforce them, removing rubble, moving people, especially the vulnerable, to higher floors of their buildings. Still others are involved in ensuring that the early warning systems are operational, that water drainage systems are functional, that waste is disposed of, and virtually anything else their exceptional advantages, referred to earlier, allow them to take on.

In the Emergency Phase, they and their unique resources are put into action leading, giving command and control and implementing the emergency decisions taken higher up the chain of command. Military personnel join in but more often take charge of efforts as diverse as transporting evacuees and the injured, caring for the injured, building shelters, and ensuring cooking facilities; food and even pet food are available in safe centers to which evacuees are moved. This too provides security for the appliances and other valuables evacuees are obliged to leave behind them when they are removed from their exposed locations, reinforcing buildings and especially roofs, and generally moving quickly and in the midst of the disaster to mitigate its effects.[6]

In the Recovery Phase as well, the military are active although increasingly, as things become calmer and less urgent, the Party and other agencies take over and military command and control become less central to success. Nonetheless, the military can often still be seen providing accommodation, transport, logistics, and other support to the relief effort. Indeed, they are often directly involved, even at this stage, in activities as diverse as repairing and rebuilding homes, placing new roofs, producing building materials, ensuring clean water availability, supporting the work of local medical teams with military doctors and even veterinarians, returning evacuees to their homes, keeping tabs on appliances and other household goods that may have gone missing or been damaged during the emergency. The result has been, as DC is proud to boast, that the armed forces "participate in all protective actions for the population and national economic resources in facing disasters," and they do so, "with a high degree of appreciation for their labor on the part of the Cuban people" (Cobas Dávila 2014, 45). There are few if any roles of the armed forces which garner them such popular support as that of natural disaster management.

RECENT AND NOT SO RECENT TRENDS

It should be noted that the early 1980s changed much in this system and guided its development along certain new lines. With the arrival of the Reagan government in Washington, and the increasing inability of the Soviet Union to engage in a serious competition with the United States in either the economic or military sphere, Moscow let the Cuban government know that while it would always be willing to give Havana moral support in case of a crisis between Cuba and the United States, its assistance would not go beyond that.

Cuba would have to fend for itself in defense areas, as never before over the previous two decades. Fidel and Raúl took up the challenge, and one of the most important moves was the creation of an even more vast reserve system, the Territorial Militia Troops (Milicias de Tropas Territoriales, or MTT), an enormous undertaking to increase by hundreds of thousands the total manpower available to the nation from its reserve forces. The scheme involved the bringing into the defense system of many people who would not previously have been considered apt for military service of any kind. Facing this new context, Raúl saw the need as one of creating a "powerful reserve capable of multiple combat missions, as well as others related to the safety and defense of factories and installations, bridges, railroads, access routes [with] vast numbers of persons who do not enjoy the physical condition to submit to [full military] training [. . .] but who can cooperate in a multitude of tasks" (Castro 1982). The regular force would begin a decline from which it was not to recover until well into the twenty-first century, but the reserves from which the bulk of assistance to DC would come to grow by many thousands of men and women.

This was just as well. In the 2006 hurricane season, a major hurricane struck the West of the island just as the bi-annual highly important *Bastión* 2006 exercise was gearing up in the Oriente. The forces were caught rather flat-footed and in the context of the terrible financial crisis the country was experiencing, not to mention the enormous cuts in money, manpower, equipment, training, and exercises the FAR had been facing by then for a decade-and-a-half, the military response was relatively slow although in the end quite effective. But the rarely criticized armed forces did come in for some significant complaints, something that stung them badly. It would not happen a second time. And improvements to, and resources for, the system became a greater priority. The MTT's growth helped here.

There has also been a growing international context for Cuba's Civil Defense system. Lauded by the United Nations and many other multilateral and national organizations, a number of countries have studied the Cuban approach to the phenomenon. Praise from, and cooperation with, the UN has been especially impressive including a visit by then Secretary-General Ban Ki-Moon to Cuba's Civil Defense Headquarters in January 2013. Indeed, it has become a frequent arrangement that during an actual hurricane or other disaster, the Venezuelan armed forces send a detachment, termed a brigade but of course not of any such size or configuration to Cuba to watch how the Cubans operate. It is not certain how much the Venezuelan context allows for actually implementing changes along Cuban lines but the interest is there. And Cuba's experience in the field has assisted its simply exceptional engagement in, and successes with, internationalist assistance to other countries around the world struck by natural disasters (Kirk 2016).

Even the United States, in the context of the *apertura* in bilateral relations of the last Obama years moved toward cooperation in the public health sphere, especially which related to natural disasters. A joint working group was established and U.S.–Cuban cooperation in West Africa against Ebola and in Haiti in the post-2010 earthquake context was notable and groundbreaking if still low key (Alzugaray Treto 2017). The United States even proposed that Cuba be invited to attend as an observer the Conference of Ministers of Defense of the Americas in Santa Cruz, Bolivia, in 2010 and to speak to the ministers about how Cuba handled natural disasters.

Cuba has also hosted since 1987 a series of international conferences dealing with natural disaster management, organized by the National General Staff of DC, under the auspices of the United Nations Development Programme. In these, it has learned from other countries, but given its exceptional record in this field, it has tended to be the nation sharing its good practices with others, especially those in the Caribbean area and in Latin America. These conferences have stimulated great interest among authorities, functionaries, scientists, professionals, specialists, and activists from around the world dealing with disasters.[7]

CONCLUSIONS

"First in, last out." This expression, usually used for special forces in a variety of countries, can well apply to the Cuban armed forces where natural disasters are concerned. They are at the forefront of natural disaster risk management on a permanent basis and when an emergency strikes; their unique capacities ensure they have the lion's share in the leadership of the preparation, immediate emergency actions, and even to a great extent eventual recovery in the face of such situations.

They are indeed the first and last bulwark of the nation's response to natural disasters, a role they take seriously and fulfill, by all reports, very well indeed. No other organization can compare with them in terms of capabilities in emergencies, and it has fallen naturally to them to act as a coordinator and a major player, in dealing with natural disasters. But they are the first to say that it is only through their cooperation with mass organizations of the nation, the Party, the reserves of the FAR themselves, MININT, and the vast number of local authorities, that progress is made and damage and loss of life are limited in the exceptional way it tends to be in Cuba when disasters strike. As usual in Cuba, they may be *primus inter pares,* but they are not the whole story. The nation as a whole, organized to respond to defense against aggression or natural disasters, is the key to success. It is that public, ordinary people educated and even trained for the job, and having a culture of mobilization and social organization, that is paramount. The FAR are, as so often in Cuban national life, merely the central pillar in making both jobs turn out so successfully.

NOTES

1. The term "pseudo-colonial" is used in Cuba to designate the period between the end of Spanish colonial rule in 1898 and the beginning of an informal U.S. protectorate over the island, and the "triumph" of the Castro Revolution on 1 January 1959.

2. This is the official list given by Civil Defense in Cuba. The phenomenon of tornadoes, so destructive recently in Cuba, should be added to this list.

3. The author is able to elaborate on these points in his *Raúl Castro and Cuba: A Military Story*. 2011. New York: Palgrave Macmillan, 16–18.

4. In Cuba, this date is celebrated as the birth of Civil Defense, the national system dealing with natural disasters.

5. This figure does not, of course, include the EJT, itself numbering 100,00 because they are not in any real sense part of the permanent forces, working exclusively in agricultural tasks and having undergone very little strictly military training.

6. This list is far from exhaustive. One is astounded by the range of things the military and other agencies coordinated by them engage in during these first two phases. These few come from conversations with Civil Defense officers conducted in Cuba between 2004 and 2010.

7. Related to this, Cuba has also hosted a series of international conferences for fire brigades.

BIBLIOGRAPHY

Álvarez Calvo, Jorge Luis. 2015. *Enfrentamientos a Desastres*. Havana: Editorial ACPA.

Alzugaray Treto, Carlos. 2017. "Cuba's National Security vis-à-vis the United States: Conflict or Cooperation?" In *Debating U.S.–Cuban Relations: How Should We Now Play Ball?*, edited by Jorge I. Domínguez, Rafael M. Hernández and Lorena G. Barberia, 62–81. New York: Routledge.

Castro, Raúl. 1973. "Discurso del V Aniversario de la Columna Juvenil del Centenario y la Constitución del Ejército Juvenil de Trabajo." In *Discursos y Artículos*, 245–259. Havana.

Castro, Raúl. 1982. "La Constitución de un batallón masculina y una compañía femenina de las Milicias de Tropas Territoriales," Santiago, January 21. In *Discursos y artículos,* Vol. II, 169–180. Havana.

Cobas Dávila, Lt-Col. Wilfredo. 2014. "Las Fuerzas Armadas Revolucionarias y el Ministerio del Interior, Junto al Pueblo en Situaciones de Desastres." In *Defensa Civil de Cuba*, Vol. I, 44–45. Havana.

Defensa Civil. 2017. *La Defensa Civil Cubana*. Havana: Casa Editorial Verde Olivo.

del Pilar Díaz Castañón, María. 2001. *Ideología y Revolución: Cuba 1959–1962*. Havana: Ciencias Sociales.

Díez Acosta, Tomás. 2008. *Octubre 1962: a un paso del holocaust*. Havana: Editora Política.

Etcheverry Vázquez, Pedro, and Santiago Gutiérrez Oceguerra. 2008. *Bandidismo: derrota de la CIA en Cuba*. Havana: Editorial Capitán San Luis.

Forteza, Francisco. 2003. "Cuba: Ejército productive." World Data Service, WDS-015. August 5.

Francisco Álvarez, Juan. 1994.*Cuba sesenta siglos antes de Colón*. Havana: Editorial Publicigraf.

García Moreno, María Luisa. 2011. *Historia Militar de Cuba*, Vol. II. Havana: Editorial Verde Olivo.

Herrera Medina, José. 2006. *Operación Jaula: contragolpe en el Escambray*. Havana: Editora Verde Olivo.

Kirk, John M. 2016. *Salud Pública sin fronteras*. Santiago de Cuba: Editorial Oriente.

Klepak, Hal P. 2011. *Raúl Castro and Cuba: A Military Story*. New York: Palgrave Macmillan.

Kornbluh, Peter, editor. 1998. *Bay of Pigs Declassified: The Secret CIA Report on the Invasion of Cuba*. New York: New Press.

Lezcano Pérez, Jorge. 2008. "La Defensa de la Revolución por las masas." In *Memorias de la Revolución*, Vol. II, 132–140, edited by Enrique Oltuski Ozacki, Héctor Rodríguez Llompart and Eduardo Torres-Cuevas. Havana: Editora Imagen Contemporánea.

Malmierca, Maris Teresa. 2005. *Comités de Defensa de la Revolución: aporte de la Revolución cubana*. Havana: Ediciones Extramuros.

Millán Álvarez, Maira. 2014. "Gestión del riesgo de desastres con enfoque de género." In *Defensa Civil de Cuba*, Vol. I, 16–19. Havana.

Morley, Morris H. 2002. *Imperial State and Revolution: The United States and Cuba, 1952–1986*. Cambridge: Cambridge University Press.

Nguyen Giap, General Vo. 2008. *Guerra del pueblo, ejército del pueblo*. Havana: Ciencias Sociales.

Núñez Jiménez, Antonio. 1982. *El Archipiélago*. Havana: Editorial Letras Cubanas.

Paterson, Thomas G. 1994. *Contesting Castro: The United States and the Triumph of the Cuban Revolution*. New York: Oxford University Press.

Pavlov, Yuri. 1993. *The Soviet-Cuban Alliance 1959–1991*. New Brunswick, NJ: Transaction Publishers.

Puig González, Miguel Ángel, José Ernesto Betancourt Lavastido, and Rolando Álvarez Cadeño. 2010. *Fortalezas frente a huracanes (1959–2008)*. Havana: Editorial Científico-Técnica.

Rueda Jomarrón, Hugo. 2009. *Tradiciones combativas de un pueblo: las milicias cubanas*. Havana: Editora política.

Suárez Rivas, Ronal. 2014. "Isabel Rubio y las inundaciones: un récord que nadie desearía tener." In *Defensa Civil de Cuba*, Vol. I, 23–26. Havana.

Waters, Mary Alice. 2003. *Marianas in Combat: Teté Puebla and the Mariana Grajales Platoon in Cuba's Revolutionary War 1956–58*. New York: Pathfinder.

Chapter 4

Cuba–Russia Cooperation

The History of Fraternal Disaster Management Collaboration

Isabel Story

Cuba is an international paradigm for successful disaster preparation and management. This international reputation has come to the fore particularly since the fall of the Soviet Union. However, as a number of the authors in this book highlight, since the earliest days of the revolutionary process, significant emphasis has been placed on minimizing the loss of life and mitigating the damage caused by natural disasters, particularly after the devastation of Hurricane Flora in 1963 (Popausek and Spata 1964). Education and research have also been fundamental pillars of the country's response to the threats posed by natural disasters (Llanes Guerra 2008). The Soviet Union played an integral role in the development of Cuba's excellent disaster preparedness capabilities, principally in the field of collaborative scientific research. This chapter will briefly trace the evolution of Cuban–Soviet relations and will predominantly focus on joint collaborations in the Caribbean, linking developments in research collaboration back to wider trends and tensions in their alliance.

THE CUBAN–SOVIET ALLIANCE: AN OVERVIEW

The Cuban–Soviet alliance was one borne out of necessity rather than a genuine ideological impetus. The long and oft-thwarted project for national independence finally came to fruition in the midst of the Cold War, which quickly shaped the course of the revolutionary project. Diplomatic ties between Cuba and the Soviet Union had been cut in 1952, as Fulgencio Batista clearly sought to ally himself with the United States. However, they were restored in 1960, against the backdrop of a cautious initial agreement to purchase a relatively small amount of Cuban sugar the year before. As the rapid decline of U.S.–Cuban relations made finding a new trading partner an urgent task, the Cuban–Soviet relationship developed (Medvedev 1965). In 1962, the Cuban Missile Crisis, also known as the October Crisis, brought the Cold War firmly to Latin America, creating a new epicenter of international tensions. However, the fiasco also provided Cuba with unprecedented stability and economic support, owing to the agreements that it managed to negotiate with the USSR in the wake of the event.

As Cuba became increasingly geopolitically isolated, the Soviet Union was one of the few options that the burgeoning Revolution had in order to obtain knowledge and goods. Economic aid quickly spiraled, and the two countries found themselves in a relationship from which it was difficult to extricate themselves, described by LeoGrande and Thomas (2002, 342) as "golden handcuffs." Cuba needed the stability and protection that the alliance with the Soviet Union provided to pursue its ambitious nation-building plan and support of other national independence movements. And the USSR needed the prestige among the international socialist camp and the dynamism that the success of the Cuban Revolution offered. Thus, Cuba became a full member of the Council for Mutual Economic Aid (CMEA, also known as COMECON) in 1972, and trade between the two countries accounted for an increasingly large percentage of Cuba's economy (Hoffmann 1977). The year 1985 marked the high point in trade between the two countries at almost 10 billion Rubles (Bain 2005, 774). The same year, the Soviet Union provided huge amounts of humanitarian aid after Hurricane Kate. However, alongside the continuing—seemingly ever-growing—economic assistance from the USSR in the form of advantageous trade deals, Cuba continued to have to focus on sugar production, causing uncomfortable parallels with its prerevolutionary economic history. The spike in world oil prices in 1978 pushed Cuba into an even closer relationship with the Soviet Union, which became increasingly problematic from around 1987 as Mikhail Gorbachev's policies of *perestroika* and *glasnost* began to take hold in the Soviet Ministry of Foreign Trade. As a result, Soviet trade with Cuba began to decline steadily.

Ultimately, the three-decade-long alliance was fraught with tensions throughout its existence. Essentially, these tensions boiled down to debates about the meanings of socialism and communism, theory versus practice and lived experience, and ideas regarding the leadership of the international socialist camp. From the mid-1960s, Cuba positioned itself as the natural leader of the global south and associated liberation movements, while the USSR espoused a more peaceful approach toward international relations. Underneath the banner of an "unbreakable friendship" and fraternal alliance, these multiple tensions played themselves out in fields where there was more discursive space, such as culture. While Cuba grew its educational capacity, beginning at the most basic level with the 1961 Literacy Campaign and then working up, many Cubans went to the Soviet Union to study at a postgraduate level.

As the immediate educational challenges were met, the focus shifted to one of greater organization and a deepening of research capacity, and a focus on achieving and maintaining economic independence. This move toward sustained and systematic research and improved organization emerged at a time when Cuban–Soviet relations were at their economically closest but increasingly strained ideologically (1975–1986). It was during this stage that the majority of Cuban–Soviet research collaboration into natural disasters occurred.

FRATERNAL AND INTERNATIONALIST HELP

Initially, reflecting the deeply pragmatic nature of early Cuban–Soviet relations, the Soviet Union's interactions with Cuba in the sphere of disaster preparedness and management was principally economic. This had begun with the donation of a prefabricated housing factory in 1964 after Hurricane Flora; the USSR had donated the same type of factory to Yugoslavia after the 1963 Skopje earthquake (Alonso and Palmarola 2014). The next hurricane to cause major damage to Cuba was Hurricane Alberto in 1982. In response to the damage caused, the USSR sent seven tons of medicine to help the victims of the hurricane and the extensive flooding it had caused, and also to be used by the Ministry of Public Health (Ministerio de Salud Pública, or MINSAP) more generally (*Granma* 1982b). Such aid was generally geared toward the global south (then referred to as the "Third World"), but in the case of Cuba, it was a "gesture of solidarity which reveal[ed] the unbreakable unity between the homeland of Lenin and the Cuban people" (*Granma* 1982b, 3). Further support, initiated by the USSR, was given in 1985 after the particularly damaging Hurricane Kate—the natural disaster which had most affected Cuba's economy (*Granma* 1985b; Barrio Menéndez 1985). This support addressed immediate food issues but was also aimed at helping mitigate the damages to the Cuban economy and helping planned work continue (*Granma* 1985a). This "fraternal and internationalist gesture" was celebrated by the press (Cardosa Yaser 1985, 1). Among other items, the aid comprised 30,000 tons of rice, 20,000 tons of wheat flour, 10,000 zinc sheets, 5,000 aluminum sheets, 2 million asbestos cement tiles, and 1,000 tons of large-fiber asbestos (*Granma* 1985a; Guma 1985).

The shipments and progress of this significant amount of aid were eagerly followed in the Cuban press (*Granma* 1986; 1986a; 1986b; 1986c; 1986d).

In the subsequent discussion of the Soviet aid, *Granma* reminded its readers of the Soviet Union's established history of sending help when Cuba was confronted with shortages and economic problems occasioned by the tense U.S.–Cuban relationship; for example, sending fuel in 1960, buying sugar when quotas were broken, or supplying goods (oil, wheat, foodstuffs, fabric, machines, prime materials) when the embargo was enacted and military support when defense issues arose (*Granma* 1985b; Peláez 1987a; 1987b). However, between these periods of major humanitarian aid, a more sustainable system of research collaboration into a shared problem had begun.

A SHARED PHENOMENON

Prior to Flora institutions that catered toward extreme weather events had begun to be established in response to the shifting needs of the Revolution as and when they emerged. This was reflective of the general nature of the revolutionary nation-building project in the first decade of the Revolution. One such institution was the Institute of Hydraulic Resources, which was founded in August 1962 in response to the significant droughts that had affected Cuba between 1960 and 1962 (Schwartz 2015). However, while steps had begun to be taken before Flora devastated the island, the catastrophe provided the impetus to develop a comprehensive disaster preparedness system, capable of responding to events, such as hurricanes, with the aim of maximum preservation of life. Existing institutions and programs were also adapted to better contribute to the developing disaster preparedness system. For example, the Popular Defense (Defensa Popular) was reorganized as the Civil Defense (Defensa Civil, or DC). Its focus was broadened to include mobilization in order to protect the population against natural disasters and/or mitigate their impact on everyday life, as well as vigilance against military threats. By the end of the 1960s, the DC had come to occupy an important role in disaster preparedness management.

In addition to efforts focused on the organization of bodies and training geared toward the mitigation of fatalities caused by natural disasters, there was also a concerted effort to improve knowledge and education about hurricanes and other natural disasters to which Cuba was prone. This educational drive took many forms, both overt and subtle. For example, the dissemination of information about ways to mitigate the destructive effects of hurricanes found its way into the 1965 publication, *Geography of Cuba*, which reproduced a fifteen-point plan from the Department of Meteorology at the Central University of Las Villas on how to help reduce the negative impact of cyclones (Núñez Jiménez 1965). Research into hurricanes also continued apace, and in 1965, the Cuban Academy of Science's Institute of

Meteorology (Instituto de Meteorología, or INSMET) was founded, comprising thirteen departments, all geared toward generating and disseminating meteorological knowledge. One of these departments was that of "Tropical Cyclology." Among its diverse roles (from organizing archives to putting every hurricane since 1800 into chronological order), it was duty-bound to publish research it conducted on hurricanes and hold seminars on the prediction of hurricanes (Academia de Ciencias de Cuba 1970).

In 1975, at the First Congress of the Cuban Communist Party (Partido Comunista de Cuba, or PCC), Fidel Castro celebrated the advances made in the field of science, including the 1965 creation of the Institute of Meteorology, which had subsequently "installed a network of modern radars and stations, as well as artificial tracking satellites that predict hurricanes in good time, helping to save lives and material wealth" (PCC 1975, 93). In keeping with the general focus on raising the education and specialist, technical knowledge levels in Cuba in the 1970s and 1980s (as articulated at the First Congress of the PCC), improving Cuba's preparedness and knowledge of hurricanes was highlighted as a significant area of needed improvement. Over the next two decades, the government continued to look for ways to improve its disaster preparation approach. As much of the population as possible was involved in the search for collective solutions, which valued collective problems. A, perhaps unexpected, element of the focus on collective solution-finding was the significant collaboration with the Soviet Union.

Like Cuba, the former Soviet Union was also subject to extreme weather conditions with increasing frequency. This was particularly true in the Far-East and the seas of the Far-East that were affected (and still are) by tropical hurricanes that form in the North-West Pacific (Krokhin 2017). There was a dedicated research body—the USSR State Committee for Hydrometeorology—with five research institutions, which were mainly concerned with tropical cyclone problems in the Soviet Union. These institutes were: The Institute of Experimental Meteorology (Obninsk), the USSR Hydrometeorological Research Center (Moscow), the Central Aerological Observatory (Dolgoprudnyi), the Far-East Regional Research Hydrometeorological Institute (Vladivostok), and the All-Union Research Institute of Hydrometeorological Information—the World Data Center-B (Obninsk) (Falkovich et al. 1990).

Significant progress in the modeling and analysis of hurricanes had been made since the 1960s (Falkovich et al. 1990). Moreover, the USSR had been active in significant international collaborative research exercises with institutes, such as the World Meteorological Organization (WMO) and the International Council of Scientific Unions (ICSU) and their long-term Global Atmospheric Research Program (GARP). The Global Atmospheric Research Program Atlantic Tropical Experiment (GATE), a major regional experiment of GARP, was carried out in the summer of 1974 under the aegis of the World Meteorological Organization and the International Council of Scientific Unions. GATE involved seventy-two participating nations (including Cuba) and took place over approximately 100 days in the summer of 1974. The goal of the exercise was to study the impact that smaller scale tropical weather systems had on large-scale motions and to improve numerical modeling and prediction methods as a result (Kuettner 1974). Such international collaboration continued throughout the 1970s, and under the sponsorship of the World Meteorological Organization in 1977 a Hurricane Committee, created in Mexico that brought together international specialists met in order to discuss regional operations (Schwartz 2015).

The growth in Soviet participation in global projects—and not just within the socialist camp—was indicative of Brezhnev's predilection for greater international co-operation (Suri 2006). This increased favoring of international cooperation echoed the renegotiation of the idea of the future that was occurring; from one that was rooted in demonstrating the superiority of the Soviet system to one that was based on protecting a "vulnerable habitat shared by all of mankind which was to be safeguarded in a common effort" (Guth 2015, 373).

This renewed focus on international cooperation, shared problems, and global modeling had led to the founding of dedicated institutes, such as the Institute of Applied Systems Analysis (IIASA) in 1972, and the Academy of Science's Institute for Systems Research (VNIISI) in 1976. The former worked on topics such as "automated control of industrial production, urban and regional planning, energy systems, ecological modelling and control, water resources, artificial intelligence, and long-term global simulation studies" (Gush 2015, 373). The latter's main task was "inspired by the work of the US Commission for the Year 2000, and consisted in forecasting the demographic, economic and environmental development of the world system and the Soviet Union" supported by the Academy of Science's Computer Centre which "specialized in forecasting anthropogenic interference with the bio- and geosphere, including the climate impact of CO_2 emissions and a possible nuclear war" (Guth 2015, 373).

Thus, the demonstrable interest of Soviet scientists and authorities in tropical cyclones was "associated not only with the efforts directed towards studying the general circulation of the atmosphere, working out medium-range weather forecasting techniques and developing climatic models, but also with direct needs of the national economy" (Falkovich et al. 1990, 3). Fraternal solidarity and economic pragmatism found common ground. For example, greater study into the formation and behavior of tropical cyclones could potentially yield results that could be used to protect and mitigate damage to the coastal and inland areas of the USSR that were sometimes adversely impacted by destructive tropical cyclones. However, it also reflected the new way of visualizing the future, and the overwhelming focus on bureaucratic efficiency of the Brezhnev administration. This focus was centered on two theories, including Scientific-Technical Revolution, (nauchno-tekhnicheskaia revolutsiia, or STR) and Scientific Management of Society (nauchnoe upravlenie obshchestva, or SMS), minor ideological innovations characteristic of the increasingly immobile Soviet political system (Brudny 2000). The STR was considered an essential prerequisite for the building of communism (Guth 2015) but was also an inherently more discursive and flexible idea and term than "communism" over which the Communist Party of the Soviet Union held a near-total claim. The STR was envisioned as a "consultative process in which political leadership was to be joined by technocratic expertise—invited discussion by a broad range of actors and was thus open to negotiation to a completely different degree than the term 'communism'" (Guth 2015, 355). The potentially more flexible and accommodating nature of the term that increasingly governed international interactions was particularly useful for the Cuban–Soviet relationship, which was frequently characterized by tensions brought about by different conceptions of communism.

As part of the STR and the greater focus on organizational clarity within the Cuban Revolution, there was increased coordination of the Cuban–Soviet relationship through dedicated institutional structures, such as the Intergovernmental Soviet–Cuban Commission for Economic, Scientific, and Technological Cooperation, which had been running since 1970. In the second half of the 1970s, the collaboration between Cuba and the Soviet Union broadened to include the study of tropical meteorology. From 1980 onward, Vietnam was also an active partner with the Soviet Union in research into tropical meteorology (Falkovich et al. 1990). Agreements on scientific and technical cooperation in the field of tropical meteorology and hurricane research were signed between the USSR and Cuba in April 1977. As a result of this agreement, a Joint Soviet–Cuban Laboratory (JSCL) was established and began its activities in Havana in 1979. This was shortly followed by the establishment of a Joint Soviet–Vietnamese Laboratory (JSVL), which began its activities in Hanoi in 1982 (Falkovich et al. 1990). The JSCL had a permanent staff of four to five Soviet specialists and approximately 30 Cuban specialists. The establishment of these well-equipped, permanent, laboratories represented a significant step forward in the possibilities for sustained study of tropical cyclones. Both the JSCL and the JSVL had four main goals in their research (Falkovich et al. 1990, 8):

- to obtain more detailed knowledge about the nature of tropical cyclones;
- to develop the system of observations on tropical cyclones;

- to improve the forecasts of tropical cyclone formation, evolution, and motion, with due consideration for regional features;
- and to create a data bank on tropical cyclones using historical and contemporary data for use in further research of operations.

Building on the 1977 convene, in 1979 a new set of agreements were signed between Cuba and the Soviet Union, to come into effect from 1980. The signing of this agreement was celebrated as a demonstration of solidarity between Cuba and the Soviet Union but also as a landmark collaboration between a "developed" country and a Latin American country in the study of hurricanes, the rains caused by hurricanes, and other aspects related to tropical meteorology (Osa 1979b, 5). Over the next decade, a number of significant research projects were carried out by the international team, including marine, aerial, and land expeditions (Falkovich et al. 1990). For example, two marine expeditions aimed at studying the characteristics of tropical cyclones and background atmospheric conditions in the Caribbean Sea were successfully conducted in 1979 and 1981; two special-purpose expeditions (Typhoon 1975, 1978) were conducted to find out more information about tropical cyclone energy and structure (Ibid.).

The continued cooperation between Cuba, the Soviet Union, and allies from the global south did not pass unnoticed in the public press and *Granma* regularly ran stories reporting on the progress of joint collaborative efforts. In 1979, the upcoming joint research venture made front-page news. *Granma* reported that the research boat "Professor Zubov," named after the founder of the Institute of Oceanology in the USSR (Osa 1979a), had set out on its 26th research trip with a team of sixty Soviet scientists and seven Cuban specialists from INSMET. The trip, which was part of an agreement signed between the State Committee for Hydrometeorology of the Soviet Union and the Cuban Academy of Sciences, aimed to learn more about the physical mechanisms involved in the formation of tropical hurricanes (Coro 1979). The research trip also brought the international socialist community together more widely: the boat had been built in the GDR following a project of Soviet naval engineers, and this was its first outing in the Caribbean Sea (Ibid.). Accompanying the team on the boat was Victor Zaria, the Cuban representative of the State Committee for Science and Technology of the USSR. The following day, *Granma* dedicated a full page spread to the trip, which was to last approximately one month and address a number of key questions that would help clarify why and how hurricanes are formed (Osa 1979a).

The publication emphasized the human aspect of the research, highlighting the youth of the Cuban team that had an average age of 30. Two of the group had already participated in Cuban–Soviet research ventures, "Tropex-74," the acronym for the North Atlantic Tropical Experiment and the already mentioned "Typhoon-78." In discussing their participation in this and the previous exhibitions, the team members noted the opportunity this fraternal venture presented for Cuban scientists to learn from the knowledge and experience of the Soviet scientists (Osa 1979a). The trip was a success, collecting a significant amount of data that was to be processed in Leningrad before being shared with Cuba's INSMET (Osa 1979c). The trip was also celebrated as being just the start of the joint research ventures, which would continue in boats and "hurricane-hunting" planes (Ibid).

Throughout the remainder of the 1980s, Cuban–Soviet research collaboration blossomed, and *Granma* continued to chart and celebrate the progress made as a result of the alliance. A follow-up research trip from the 1979 expedition was conducted in 1982, when ten Cuban meteorologists joined Soviet scientists aboard the "Academic Koroliov" boat to further study the conditions that caused cyclones to form (*Granma* 1982a). From 1983 onward, the Cuban–Soviet collaboration expanded its focus and reach, and regular studies were conducted within the joint Soviet–Cuban and Soviet–Vietnamese expedition (Falkovich et al. 1990). From 1984, the JSCL began to present its findings at international research events (Díaz 1987; *Granma* 1984; Peláez 1987a, 1987b).

Scientific progress continued apace in the late 1980s, with complex land expeditions in 1984 and 1987 aimed at gathering further data on the formation and growth of hurricanes (Falkovich et al. 1990). With the signing of a new agreement in 1986, aerial laboratory work also began (Beryulev 2009), and a plane laboratory was swiftly put to use with the advent of Hurricane Emily (García Suarez 1987; *Granma* 1987a, 1987b). Perhaps the most dramatic of these aerial collaborations occurred from September 11 to 15, 1988, when five missions were carried out in the core area of Hurricane Gilbert, which passed 200–300 kilometers South of Cuba (Falkovich et al. 1990). It was a particularly powerful tropical cyclone, as described by Rappaport and McAdie (1991, 1) in a report for the United States' National Hurricane Center: "Hurricane Gilbert reached the lowest pressure (888 mb) ever recorded in an Atlantic tropical cyclone. This pressure also currently stands as the lowest sea-level pressure on record in the Western Hemisphere."

The missions conducted by Soviet–Cuban scientists provided operational data on Gilbert, which were used by the Storm Warning Service of the Republic of Cuba when taking measures designed to alleviate the damage caused by the hurricane (Falkovich et al. 1990). Each day the flights transmitted the exact position of the hurricane's center and its principal characteristics back to Havana thereby helping to mitigate damages caused (Beryulev 2009, 63). However, after the successes of the Hurricane Gilbert flights, Cuban–Soviet cooperation slipped out of the public eye and little more was reported on it, as the Soviet Union began to disintegrate, and Cuba entered the Special Period.

CONCLUSION

The development and achievements of the JSCL reflect the evolving Cuban–Soviet relationship and its nuances. Initially a one-directional relationship, Cuban–Soviet interactions evolved into a more dynamic and multifaceted alliance from which both parties benefited in their shared goal but also helped them to achieve their individual aims. Early interactions provided a clear, public affirmation of support and delivered humanitarian aid that was genuinely in need (Díaz-Briquets and Pérez-López 2000). However, it also allowed the revolutionary government in Cuba to continue to pursue the goals of the Revolution and address pressing issues that had been further exacerbated by hurricanes (such as housing or ambitious building targets) and in doing so helped present the Soviet Union in a positive, although potentially paternal, role. As the Cuban Revolution entered its second phase and the emphasis moved from mobilization to modernization, the alliance became increasingly coordinated at institutional levels and managed through specific institutes accordingly. The organization and channeling of interactions reflected the increased focus on seeking answers to specific questions that were deemed to be important to the development and resilience of the Revolution, and inclusion of the Cubans in the search for these answers. This new organizational approach also served the Soviet Union's specific goals of protecting the planet and the future, and thus helping to demonstrate Soviet superiority, and also the building of communism through science and innovation.

Ultimately, as in other areas of society, education and research collaboration with the Soviet Union was a way for Cubans to assert themselves as equals within a multi-layered and evolving relationship with a partner whose global profile often threatened to eclipse Cuba. In doing so, this collaboration yielded real, tangible results that have actively contributed to the development of shared knowledge about the formation and behavior of hurricanes. The commitment to joint research performed two other important functions, including helping to advance shared and individual goals, as well as acting as an international affirmation of the unity of the international socialist camp.

BIBLIOGRAPHY

Alonso, Pedro Ignacio, and Hugo Palmarola. 2014. *Panel*. London: Architectural Association Publications

Bain, Mervyn J. 2005. "Cuba-Soviet Relations in the Gorbachev Era." *Journal of Latin American Studies* 37 (4): 769–791.

Barrio Menéndez, Emilio del. 1985. "Encomiable disposición de portuarios." *Granma*. December 5: 2.

Beryulev, G.P. 2009. "Scientific Research Flights into the superhurricane Gilbert (1988)." *Russian Meteorology and Hydrology* 34 (1): 62–64.

Brudny, Yitzhak M. 2000. *Reinventing Russia: Russian nationalism and the Soviet state, 1953–1991*. Cambridge, MA: Harvard University Press.Bottom of Form

Cardosa Yaser, Dantes. 1985. "El pueblo cubano agradece la fraternal ayuda de la Union Sovietica." *Granma*. November 29: 1.

Coro, Arnaldo. 1979. "Hacia el Caribe expedición soviético-cubana para estudiar los huracanes tropicales." *Granma*. September 20: 1.

Díaz, Orfilio. 1987. "Orfilio Díaz, Primer Congreso Internacional sobre Desastres Naturales." *Granma*. January 15: 2.

Díaz-Briquets, Sergio, and Jorge F Pérez-López. 2000. *Conquering Nature: The Environmental Legacy of Socialism in Cuba*. Pittsburgh, PA: University of Pittsburgh Press.

Falkovich. A. I., L. Kh. Ingel, V.P. Karmazin and Alexander Khain et al. 1990. *Tropical Cyclones: Their Studies in the Soviet Union and Some Results*. Leningrad: Leningrad Gidrometeoizdat.

Granma. 1982a. "Cubanos en expedición científica." January 18: 5.

Granma. 1982b. "Dona la Cruz Roja de la URSS a Cuba siete toneladas de medicamentos." June 29: 3.

Granma. 1984. "Efectuaran seminario internacional sobre desastres naturales." August 23: 3.

Granma. 1985a. "Importante ayuda de la URSS: Nota del Consejo de Estado." November 28: 1.

Granma. 1985b. "Una vez más la mano amiga y generosa." November 30: 1.

Granma. 1987a. "Llego a Camagüey avión-laboratorio soviético para continuar investigaciones sobre lluvia provocada." August 28: 1.

Granma. 1987b. "Cubanos y soviéticos penetraron en el centro de Emily." September 25: 1.

Guma, José Gabriel. 1985. "Suscriben acuerdo Cuba y la URSS." *Granma*. December 6: 6.

Guma, José Gabriel. 1986a. "Parten de la URSS dos buques con ayuda a Cuba por los danos del huracán Kate." *Granma*. January 18: 1.

Guma, José Gabriel. 1986b. "Hacia Cuba otros dos barcos soviéticos con ayuda por el huracán." *Granma*. January 25: 7.

Guma, José Gabriel. 1986c. "Virtualmente cumplida entrega de donación de la URSS a Cuba." *Granma*. May 17: 1, 8.

Guma, José Gabriel. 1986d. "Cumplida la donación hecha a Cuba por la URSS a causa del Kate." *Granma*. June 6: 6.

García Suarez, Ariel. 1987. "Ciclón: un laboratorio volante puesto al servicio de la ciencia." *Granma*. August 31: 3.

Guth, Stefan. 2015. "One Future Only. The Soviet Union in the Age of the Scientific-Technical Revolution." *Journal of Modern European History / Zeitschrift Für Moderne Europäische Geschichte / Revue D'histoire Européenne Contemporaine* 13 (3): 355–376.

Hoffmann, Erik. P. 1977. "The Scientific Management of Soviet Society." *Problems of Communism* 26 (3): 59–67.

Krokhin, V.V., A. Yu Fil, and V. Yu Veryatin. 2017. "Long-term Variations in the Frequency of Typhoons and Their Connection with Various Formation Factors." *Russian Meteorology and Hydrology* 42 (12): 766–774.

Kuettner, Joachim. 1974. "General Description and Central Program of GATE." *Bulletin of the American Meteorological Society* 55 (7): 712–719.

LeoGrande, William M., and Julie M. Thomas. 2002. "Cuba's Quest for Economic Independence." *Journal of Latin American studies* 34 (2): 325–363.

Llanes Guerra, José. 2008. "Cuba: paradigma en la reducción de riesgo de desastres." Report. Accessed September 20, 2020. https://www.preventionweb.net/files/2558_CubaParadigmaenlareducciónderiesgodedesastres.pdf.

Medvedev, L. 1965. *Friend of the Peoples*. Available through: Adam Matthew, Marlborough, Socialism on Film: The Cold War and International Propaganda. Accessed April 16, 2020. http://www.socialismonfilm.amdigital.co.uk/Documents/Details/N_508727_FRIEND_OF_THE_PEOPLES.

Osa, José de la. 1979a. "Partio rumbo al Caribe una expedición científica soviético-cubana: estudiara el origen, formación, desarrollo y posible trayectoria de los huracanes." *Granma*. September 21: 3.

Osa, José de la. 1979b. "Incrementaran la URSS y Cuba Investigaciones Conjuntas en Torno a los Huracanes Tropicales." *Granma*. October 22: 5.

Osa, José de la. 19179c. "Continuaran en Institutos de Leningrado el procesamiento de los datos recogidos en el Mar Caribe por el barco de Investigaciones Científicas 'Profesor Zubov.'" *Granma*. November 15: 3.

Peláez, Orfilio. 1987a. "Exponen métodos para el estudio de los ciclones tropicales." *Granma*. April 15: 2.

Peláez, Orfilio. 1987b. "Finalizo Simposio sobre Meteorología Tropical." *Granma*. April 18: 7.

PCC. 1975. *Informe del Comité Central del PCC al Primer Congreso*. Havana: Departamento de Orientación Revolucionaria del Comité Central del Partido Comunista de Cuba.

Popausek, Jivi, and Jan Spata. 1964. *Death was Named Flora*. Available through: Adam Matthew, Marlborough, Socialism on Film: The Cold War and International Propaganda. Accessed April 16, 2020. http://www.socialismonfilm.amdigital.co.uk/Documents/Details/N_508242_HURRICANE_FLORA.

Rappaport, Edward N., and Colin J. McAdie. 1991. "Hurricane Gilbert (1988) in Review and Perspective." National Hurricane Center, Florida. NOAA Technical Memorandum NWS NHC 45. Accessed February 20, 2020. https://www.nhc.noaa.gov/pdf/NWS-NHC-1991-45.pdf.

Schwartz, S. 2015. *Sea of Storms: A History of Hurricanes in the Greater Caribbean from Columbus to Katrina*. Princeton, NJ: Princeton University Press.

Suri, Jeremi. 2006. "The Promise and Failure of 'Developed Socialism': The Soviet 'Thaw' and the Crucible of the Prague Spring, 19641972." *Contemporary European History* 15 (2): 133–158.

Chapter 5

Bastión

The Shaping of a Pueblo Combatiente *and Natural Disaster Management*

Anna Clayfield

The idea that the defense of the Revolution should be the collective endeavor of the Cuban population has been a cornerstone of the revolutionary project since it came to power in January 1959. Having achieved victory off the back of a people's war against the Batista dictatorship, the newly crowned revolutionary leadership was keen to maintain the momentum of popular participation when it came to protecting the Revolution's hard-won achievements. In that first month in power, Che Guevara famously spoke of the need for the Rebel Army (the armed movement which had led the insurrection) to expand its ranks to include the then population of approximately 6 million Cubans, if the Revolution were to have any chance of survival (Guevara 1970). Throughout 1959, Fidel Castro would similarly leave little room for doubt that Cubans were expected to play their part in protecting everything the Revolution stood for, declaring in the November of that year: "Above all we have an indisputable duty: the duty to defend the Revolution"[1] (Castro 1959, 25).

This emphasis on the primacy of defense in the early years referred specifically to the country's military capacity for safeguarding the nascent, and therefore vulnerable, Revolution in the face of an enemy attack. As this volume demonstrates, the notion of defense has since expanded to include natural disaster preparedness as a fundamental component of protecting the Revolution's gains and, above all, its people. This two-fold focus is perhaps most clearly encapsulated by the creation of the National Defense System in 1966, which, along with numerous other concrete initiatives, has given order and structure to the island's defense efforts, whether in response to natural disasters or potential enemy aggression.

This chapter will examine one of the most prominent defense initiatives of the last six decades: The War of All the People (*Guerra de Todo el Pueblo*, or GTP), launched in 1980 in response to a re-heating of Cold War tensions. This nationwide defense strategy centered entirely on military preparedness and symbolized a renewed emphasis on mass participation in defense-related activities. In more recent years, however, one of these activities, the strategic military defense exercise Bastión, has also gradually incorporated natural disaster management and preparation. What follows aims to demonstrate how this broadening of Bastión's focus has occurred smoothly for two principal reasons: first, as a result of the overlap between the practical approaches required for both military and natural disaster-related defense; and second, thanks to a population ready and willing to defend the Revolution in any capacity.

THE *GUERRA DE TODO EL PUEBLO* AND THE LAUNCH OF BASTIÓN

With the consolidation of the "Sovietization" process and the firm establishment of the rule of the Cuban Communist Party (Partido Comunista de Cuba, or PCC), the Cuban Revolution appeared to be enjoying a period of stability as it entered its third decade in power. From 1980 onward, however, the Revolution faced a series of internal and external pressures, beginning with the Mariel Crisis of April–October 1980, a mass exodus of approximately a quarter million Cubans to the United States (Gott 2004). The inauguration of Ronald Reagan in the January of the following year did little to calm an already rattled leadership. The new president's hardline stance against Cuba and Latin America in general raised concerns of a U.S. invasion, fears which were only compounded by the presence of the U.S. Navy in the Caribbean in October 1981, the United States' invasion of Grenada in 1983, and the on-going Contra war in Nicaragua.

Sensing the very real threat of attack for the first time since the early 1960s, the response of the Cuban leadership was swift and far-reaching in the shape of the aforementioned *Guerra de Todo el Pueblo* (GTP). The initiative was based on the long-held belief that every Cuban should have a means, a form, and a place to confront an enemy invasion (EcuRed 2017). An important aspect of this strategy was the creation in May 1980 of the Territorial Militia Troops (Milicias de Tropas Territoriales, or MTT), a vast force of volunteer reservists who would assist the Revolutionary Armed Forces (Fuerzas Armadas Revolucionarias, or FAR) in defending the Revolution. By 1982, more than half a million militia members had signed up for duty (Klepak 2012).

While the GTP strategy emerged as a specific and direct response to a changing geo-political climate, its underpinning tenet—that the nation's people assume responsibility for its protection—was already firmly embedded in the Cuban consciousness by 1980. The rhetoric of the rebel leaders regarding popular defense in the lead up to and following their victory in January 1959 was soon put into practice with the creation of the National Revolutionary Militias (Milicias Nacionales Revolucionarias, or MNR) in October of the same year, a country-wide force of thousands of reservists, and the prototype for the creation of the MTT some twenty years later. MNR members frequently carried out non-combat assignments, but fully came into their own by serving alongside the FAR on the frontline during the Bay of Pigs invasion in April 1961. The defeat of the U.S.-backed invaders offered definitive proof that the combined efforts of the Cuban people could withstand a direct attack on the island's very shores.

It should be noted, however, that the idea that the popular force (i.e., the collective strength of the Cuban people) wields sufficient power to confront much larger enemies and seemingly insurmountable odds long pre-dates the guerrilla insurrection of the 1950s. The wars of independence against Spain of the late 1800s constitute a prime example of Cuban citizens of all backgrounds combining forces in the name of national sovereignty. As such, the revolutionary leadership has consistently, since 1959, called upon the image of these independentist rebels as models of citizen participation to emulate in defending the concept of the Cuban nation. The stress upon *lucha* (struggle) and *deber* (duty) as principal features of Cuban identity in the writings of the so-called "Apostle" of independence, José Martí, have accompanied this image; Cubans are repeatedly informed of their inherent duty to defend country and Revolution, as essentially one and the same thing, as a means of honoring their forebears. In its attempts to embed fully the notion of popular participation and collective action, revolutionary discourse also traces this phenomenon as far back as the early colonial period, proposing that the resistance to colonization, demonstrated by the island's indigenous inhabitants, constitutes the first example of defense involving popular forces (Clayfield 2019).

In the early 1960s, these ideas became further enshrined in speeches but also in the writings of Guevara, whose seminal work *La Guerra de Guerrillas* (Guerrilla Warfare) (1960) advocated for the potency of popular forces. The very creation of the FAR in 1959 was also grounded in this principle. Having developed out of the Rebel Army, it was, and remains, a people's army, one that was created by the people, to serve the people. The fact that many of those who wield political power are also members of the armed forces serves to blur the line between the civilian and military spheres, as does the incorporation into the FAR of thousands of civilian reservists. With this context in mind, the task of defense in revolutionary Cuba has never been the sole domain of the military, given that the latter does not exist as a separate, monolithic entity.

By the time of the launch of the GTP and the MTT in 1980, the Cuban people had long been primed to participate in preparatory defense activities and training. The GTP was not, therefore, a radically new approach to defense but rather, as the leadership describes it, a "summing up of the historical experience of the nation regarding confrontations with numerically and technologically superior enemies" (EcuRed 2017). Its principal aim was, and remains, to continue to forge a culture of a *pueblo combatiente* (combatant people), an unavoidable motif in revolutionary Cuba of the early 1980s (*Granma* 1980).

This tradition of collective defense preparation resulted in the successful completion of the inaugural Bastión[2] exercise in 1980. It put to the test all levels of the country's military structures, from the strategic to the tactical, and saw military defense officially cease to be the sole domain of the FAR, instead becoming the concern of the state and all of society, under the leadership of the PCC (Pasanaut Suárez 2014); alongside the FAR, the Ministry of the Interior (Ministerio del Interior, or MININT), bodies of the Central Administration of the State, as well as provincial branches of People's Power (Poder Popular), participated for the first time (Rodríguez Rodríguez 2016). Its enactment enabled a thorough evaluation of the organizational defense structures at national, provincial, and municipal levels, leading, in turn, to efforts to perfect the entire system (EcuRed 2019).

Bastión grew in size for its second installment in spring 1983. It consisted of the realization of a series of simulated scenarios across the island—such as an air raid—that were designed to recreate the conditions of an enemy strike as realistically as possible. Other operations comprised, among many more, extensive evacuations and people opening up their homes to fellow citizens in need, with the drive for self-sufficiency in times of adversity a central concern (La Rosa 1983a, 1983b); all of these activities mirrored those conducted as part of pre-existing natural disaster management initiatives, although Bastión's single purpose in its early years was preparation for the potential of war.

The final Bastión of the decade in 1986 turned out to be the country's largest ever military exercise to that point (García and Edith 2004), with the participation of millions of Cubans, in the guise of military personnel, *milicianos*, reservists, or civilians (Mazarr 1990, 283). Building on and expanding the activities of the previous two Bastiones, it led to the conception of Cuba's Territorial Defense System (Sistema Defensivo Territorial), one of the pillars of the 1994 National Defense Act 75 (Cuba Defensa 2014; EcuRed 2019).

The growth of the MTT and the successful completion of three Bastión exercises in the 1980s proved that the Cuban leadership was still able to mobilize the population around defense to an extent not seen since the heady days of the 1960s. As part of the wider GTP strategy, they further cemented the tradition of collective defense efforts, with defense understood solely in the context of withstanding an enemy aggression. While it would take almost another twenty years for these initiatives to include natural disaster preparation in their remit, they nonetheless revitalized the pre-existing ethos of collective action and strengthened the mass mobilization organizational structure, within which a broader concept of defense could fit neatly and seamlessly.

BASTIÓN IN THE NEW MILLENNIUM: INCORPORATING NATURAL DISASTER PREPARATION

Following the three highly successful Bastión exercises of the 1980s, such large-scale defense activities were forced into a hiatus as Cuba entered the 1990s. The collapse of the Soviet Union, Cuba's principal trading partner and ideological ally, led to the onset of the so-called "Special Period," an unprecedented economic crisis. The leadership directed all of its efforts toward rescuing the country's ailing economy, to ensure the Revolution's survival. Given the extreme hardships suffered by the Cuban population over the course of the decade, there existed neither the will nor the resources—especially since Soviet military assistance had abruptly ended—to mobilize around defense to any great extent.

Things started to look up, however, by the turn of the millennium. Heavy investment in tourism had encouraged economic growth, while other reforms, including more sustainable practices concerning food production and distribution, had enabled the Revolution to weather its most severe storm. Moreover, the Cuban people had once again proven their resilience in the face of adversity, having seemingly heeded the government's call for citizens to display unyielding *protagonismo* (assuming an active role in efforts to support the Revolution) and *voluntad* (the will to participate).

This turnaround allowed the Cuban leadership to turn its attention once again to collective defense in the early 2000s, not least because of increasing tensions in U.S.–Cuba relations. The George W. Bush administration was explicit in its intention to put an end to communist rule on the island, and, following the events of 9/11, Cuba was designated a state sponsor of terrorism (Brenner and Eisner 2017). These growing hostilities and accompanying sense of vulnerability led directly to the return of Bastión in 2004, some eighteen years after its last incarnation. Held over five days (December 13–19), it involved all of the country's Defense Councils (Consejos de Defensa) at the provincial, municipal, and local level, with the final two days designated National Defense Days, aimed at non-military, citizen-centered activities.

A high degree of popular participation was guaranteed through the incorporation of the MTT and the Production and Defense Brigades (Brigadas de Producción y Defensa, or BPD). Formed in workplaces and residential areas, the BPD, the largest of all the defense forces, offer a means of voluntary participation for those whom, for whatever reason, are unable to serve as *milicianos* or reservists (Cuba Defensa 2014); approximately 3.5 million citizens are members of more than 60,000 Brigades (Parlamento Cubano 2020). BDP's contribution to emergency situations is centered around basic defense activities and protecting economic production and services—for example, transportation, safeguarding or constructing bridges, ensuring that factories continue to operate—although members are also expected to take up arms should the emergency require it (González Abreut 2013). Established in 1984 following the experiences of the first two Bastión exercises, their formation was inspired by the groups that provided essential, non-combat support to the *mambises*, those who fought against the Spanish in the independence wars of the late nineteenth century (González Abreut 2013).

Similar to the exercises of the 1980s, Bastión 2004 focused uniquely on preparation in terms of the "real danger" of an enemy attack (García and Edith 2004); a statement issued in Cuba's oldest and most widely read general interest magazine, *Bohemia*, affirmed in the December edition that, in response to the "hostile politics" of the United States, it was necessary to continue increasing and perfecting Cuba's defense preparations under the concept of the GTP (*Bohemia* 2004). In an interview with Brigadier General Andrés González Brito in the same edition, he emphasized the underpinning, and the much-quoted, ethos of the exercise: "sweat spilled in peace is blood saved in war" (quoted in *Bohemia* 2004, 7).

Despite this emphasis on enemy aggression and war, however, many of the activities carried out during Bastión 2004 overlapped with those that form part of Cuba's long-standing commitment to natural disaster preparedness, such as during the annual *Meteoro* exercise: simulated mass evacuations, search and rescue operations, fire prevention, and the treatment of casualties in health clinics (*Granma* 2004). Moreover, all of these actions were seen to contribute to the "principal mission of preserving life, peace and the future of the Revolution" (Ibid.), itself a defining objective of Cuba's overarching approach to natural disaster management and preparation.

Thus, when natural disaster preparation was also quietly incorporated into Bastión 2009, little changed in terms of the exercise's contents and practical operation. It is unsurprising, perhaps, that, in addition to all the pre-existing disaster management initiatives, Bastión would become another opportunity for perfecting the national strategy for population preparation. Following the extensive damage caused by Hurricanes Ivan and Charley in 2004, it was recognized that the island's prevention and management system required improvement. In the December of that same year, Raúl Castro, minister of the FAR, held a meeting at the Ministry of the Revolutionary Armed Forces (Ministerio de las Fuerzas Armadas Revolucionarias, or MINFAR) headquarters, to assess deficiencies in the handling of the fallout of the two major hurricanes and to propose solutions ahead of the next hurricane season (*Bohemia* 2004). It should also be noted that the inauguration of President Obama in January 2009 relieved some of the aforementioned U.S.–Cuba political tensions. As such, the idea of collective defense encapsulated in Bastión could now safely make room for natural disaster preparation alongside its objective of protecting the nation from potential enemy attacks.

Bastión 2009, postponed by a year owing to Hurricanes Ike and Paloma in September and November of 2008, involved the participation of approximately 4 million Cubans (*Granma* 2009). Similar to the 2004 event, it incorporated many of the same tasks undertaken in relation to natural disaster management, with the unchanged principal objective of protecting human life. Participants worked to perfect strategies for protecting food production, harvesting sugar cane and other crops, safeguarding fuel and medical supplies, and carried out cleaning and sanitation exercises as part of disease control, as well as assisting with rebuilding and recovery efforts following the hurricane damage caused in 2008 (*Granma* 2009). These took place alongside air, land, and sea military maneuvers as part of simulated "imperialist" invasions (*Granma* 2009). Such maneuvers remained the driving motivation for Bastión's enactment.

For the first time, however, there was an official, if subdued and at times footnoted, recognition that preparing for both a war and/or a natural disaster required the same approach, and the latter should thus be viewed as an important component of the organization and realization of Bastión henceforth. When reflecting on the success of Bastión that year, Raúl Castro stated that, "the degree of organization achieved, which is the same whether confronting a military attack or a hurricane, is what inspires the greatest respect, more than any type of weapon" (quoted in Martín Blandino 2009). This two-pronged approach saw the front cover of *Bohemia* confidently announce three weeks after the end of the strategic exercise: "Bastión 2009: Never Unprepared!" (*Bohemia* 2009).

The desire to continue perfecting the defense system in light of a changing geopolitical climate continued to underpin the subsequent installment of Bastión in 2013. As in 2009, the event had been due to take place a year earlier but was postponed following the extensive damage caused in the Eastern provinces by Hurricane Sandy in late 2012. The thwarting of an enemy attack remained its overriding objective but, just as four years prior, reference to a concomitant preparation for natural disaster situations was also quietly woven into official rhetoric before, during, and after Bastión. Cuban news outlets placed Bastión's two central aims on an equal footing, announcing that the practical exercises undertaken by the population were designed to "fend off a military invasion or confront the event of natural disasters," while also highlighting that the fundamental tenet of protecting human life above all else continued to buttress the preparation for both scenarios (Radio Cadena Agramonte 2013). Reflecting on the success of the nationwide operation in the central provinces, including Cienfuegos and Sancti Spíritus, an article in *Bohemia* extolled the role of the country's youth in forging a defense strategy, proclaiming explicitly that "the military invulnerability of the Cuban nation is in the hands of young people" (Lezcano Lavandera 2013, 37). Such statements are consistent with the renewed emphasis on youth as holding the key to the Revolution's survival that resurged in the early 2000s with the "Battle of Ideas" ideological campaign and process of consultation and debate. That said, the same article underscored the role played by each and every Cuban—young or old, male or female—in perfecting the country's defenses. The president of the Cienfuegos Provincial Defense Council, Lidia Esther Brunet Nodarse, commented on the effective integration of participants at all levels of Cuba's defensive organizational structure, attributing this success to the general "trust and education of the people, to the conviction that each and every Cuban can be their own commander in chief" (quoted in Lezcano Lavandera 2013, 37). Notably, an estimated 1.2 million civilians took part in Bastión's closing two National Defense Days (García 2013).

The emphasis on the *protagonismo* of Cuban youth in maintaining the country's meticulously honed defensive capacity persisted in the language surrounding Bastión 2016. Younger members of the FAR were praised in the Cuban media for their dedication in directing several exercises and for applying lessons learned from their older, more experienced leaders (*Bohemia* 2016; *Granma* 2016a; *Juventud Rebelde* 2016). Again, this rhetoric chimed with a wider trend in official discourse in the 2010s for portraying FAR veterans as slowly handing over the reins of power to newer recruits, both in the military and political spheres, given that the two are invariably interlinked in revolutionary Cuba (Clayfield 2019).

In the year that marked the 60th anniversary of the landing of the Granma in 1956—the event which sparked the insurrection against Batista—the Cuban government made it clear that, despite an ever-aging leadership, defense would remain the first and foremost priority of any future leaders rising through the ranks. Moreover, the preparation for natural disasters would continue to constitute one of the cornerstones upholding the conception of defense to which the Revolution adhered: the *Guerra de Todo el Pueblo*. An article in Cuba's leading national newspaper *Granma*, for instance, described Bastión as the ultimate expression of GTP, "a doctrine that demonstrates the defensive and combative capacity achieved by the country, not only in terms of war but in the ability to confront other events such as natural disasters" (*Granma* 2016b). Regarding the latter aspect, Bastión 2016 and the National Defense Days retained the simulation of mass evacuations, search and rescue operations, and protection of goods and services as part of its focus on "armed and non-armed struggle" (*Granma* 2016c). Thus, while numerous foreign media sources reported on Bastión 2016 as a direct and immediate reaction to the election of Donald Trump to the U.S. presidency in November that year (BBC News 2016; *Clarín* 2016), it was in fact just yet another attempt to perfect and broaden a strategy conceived of in 1980. Indeed, the exercise had been officially announced by Raúl Castro in April 2016, months ahead of the U.S. election (PCC 2016).

CONCLUSION

In 1994, Cuba's National Defense Act 75 further cemented in law a principle that had long formed part of the country's Constitution: that the "defense of the socialist homeland is the greatest honor and supreme duty of every Cuban" (Cuba Defensa 2014). Mass mobilization defense exercises, such as Bastión, and in particular the two closing National Defense Days, offer a tangible means through which Cubans can put this ethos into practice. Moreover, the high level of citizen participation attained in each of the seven Bastión events is proof that the collective *voluntad* to do so is more than sufficient to secure their successful realization.

This chapter has demonstrated that, while military capacity concerning potential enemy aggression was the driver, and to a certain extent still is, behind Bastión's conception, it has over the years also been able to incorporate smoothly natural disaster preparation into its remit. This broadening of the scope of Bastión has been made possible not solely because the two types of defense often require similar practical approaches (mass evacuations, for instance), but because the embedding of the general ethos of the GTP has produced a population predisposed to understand the importance of protecting the Revolution at all costs, and to act accordingly. In other words, the efficacy of natural disaster preparation in Cuba, whether through Bastión activities, *Meteoro*, or through myriad other strategies, can be partly explained by the fact that defense preparation, regardless of the type of threat, has since 1959 been promoted by the revolutionary leadership as the domain of the *pueblo*. According to official discourse, enacting one's defense-related duties is an "act of patriotic reaffirmation" and is thus, in turn, a means of honoring the "traditions of a people who have fought for more than a hundred years to defend every inch of the homeland" (*Granma* 2004).

The most recent Bastión exercise, originally set for November 2020, was put on hold due to the COVID-19 epidemic. Yet, given that the spotlight is now placed on Cuban youth as the demographic responsible for safeguarding the Revolution and, most importantly, its people, the perfecting of defense preparation in all its forms looks set to remain a characteristic feature of the revolutionary project in the years to come. As such, the oft-repeated statement uttered by Fidel Castro in 1982, in the midst of the launch of the GTP, appears to still ring just as true nearly forty years on: "We are a people seriously determined to defend ourselves" (Castro 1982).

NOTES

1. All translations from the original Spanish are the author's own.
2. The name "Bastión" recalls the oft-repeated description in official discourse of Cuba (although it is often also attached to individual provinces and regions), as a "bastión inexpugnable" (impregnable fortress). See, for example: EcuRed (2019); *Vanguardia* (2013).

BIBLIOGRAPHY

BBC News. 2016. "Cuba Announces Military Exercises Following Trump's Win." November 9. Accessed September 2, 2020. https://www.bbc.co.uk/news/world-latin-america-37927781.

Bohemia. 2004. "Nota Informativa: Cuba es un Bastión." December 10: 37.

Bohemia. 2009. December 18, 101 (26): Front cover.

Bohemia. 2016. "Fuerza joven protagoniza Ejercicio Estratégico Bastión 2016." November 17. Accessed August 10, 2020. http://bohemia.cu/nacionales/2016/11/fuerza-joven-protagoniza-ejercicio-estrategico-bastion-2016/.

Brenner, Philip, and Peter Eisner. 2017. *Cuba Libre: A 500-Year Quest for Independence*. Lanham, MD: Rowman and Littlefield.

Castro, Fidel. 1959. "Discurso del Dr. Fidel Castro Ruz, Primer Ministro del Gobierno Revolucionario, en el acto de apertura de X Congreso de la C.T.C. (I Revolucionario)." In *Dos discursos del Comandante Fidel Castro en el X Congreso Nacional Obrero (noviembre 1959),* November, 7–40. Havana: Versión Taquigráfica de la Oficina del Primer Ministro.

Castro, Fidel. 1982. "Discurso pronunciado por Fidel Castro Ruz, Presidente de la República de Cuba, en la ceremonia militar con motivo del XXVI Aniversario del desembarco del Granma y de la fundación de las FAR, en la Plaza de la Revolución, el 11 de diciembre de 1982, 'Año 24 de la Revolución.'" Accessed July 5, 2020. http://www.cuba.cu/gobierno/discursos/1982/esp/f111282e.html.

Clarín. 2016. "Tras el triunfo de Trump, Cuba organiza un ejercicio militar de tres días contra 'acciones del enemigo.'" November 9. Accessed September 2, 2020. https://www.clarin.com/mundo/triunfo-trump-cuba-organiza-ejercicio-miltar-dias-acciones-enemigo_0_S1I-h-ZZe.html.

Clayfield, Anna. 2019. *The Guerrilla Legacy of the Cuban Revolution*. Gainesville, FL: University of Florida Press.

Cuba Defensa. 2014. "La Ley No. 75 de la Defensa Nacional." Accessed September 18, 2020. http://www.cubadefensa.cu/?q=ley75.

EcuRed. 2017. "Guerra de Todo el Pueblo en Cuba." Enciclopedia Cubana en la Red. Accessed September 18, 2020. https://www.ecured.cu/Guerra_de_todo_el_pueblo_en_Cuba.

EcuRed. 2019. "Ejercicio Estratégico Bastión." Enciclopedia Cubana en la Red. Accessed September 18, 2020. https://www.ecured.cu/Ejercicio_Estrat%C3%A9gico_Basti%C3%B3n.

García, Liset. 2013. "Ejército Oriental: Con esta tropa hay que contar." *Bohemia*. November 29, 105 (24): 39–42.

García, Liset, and Dixie Edith. 2004. "No soñamos con la guerra." *Bohemia*. December 10, 96 (25): 4–7.

González Abreut, Boris E. 2013. "Fuerza Popular." *Verde Olivo*. April, 55 (2): 49–51.

Gott, Richard. 2004. *Cuba: A New History*. New Haven, CT: Yale University Press.

Granma. 1980. "¡La Patria nos Convoca a la Marcha del Pueblo Combatiente!" April 15: 1.

Granma. 2004. "Bastión 2004: La Guerra de Todo el Pueblo." December 20. Accessed September 18, 2020. http://www.granma.cu/granmad/secciones/defensa/bastion2004/art51.html.

Granma. 2009. "Bastión 2009: La Guerra de Todo el Pueblo." November 30. Accessed September 18, 2020. http://www.granma.cu/granmad/secciones/bastion-2009/art25.html.

Granma. 2016a. "Todos por la Patria." November 16. Accessed September 2, 2020. http://www.granma.cu/cuba/2016-11-16/todos-por-la-patria-16-11-2016-16-11-43.

Granma. 2016b. "Alista el Ejército Occidental sus componentes defensivos." November 16. Accessed August 2, 2020. http://www.granma.cu/cuba/2016-11-16/alista-el-ejercito-occidental-sus-componentes-defensivos-16-11-2016-16-11-34.

Granma. 2016c. "Inicia Ejercicio Estratégico Bastión 2016." November 16. Accessed August 2, 2020. http://www.granma.cu/cuba/2016-11-16/inicia-ejercicio-estrategico-bastion-2016video.

Guevara, Ernesto. 1970. *Ernesto Che Guevara: Obras, 1957–1967*. 2 Vols. Havana: Casa de las Américas.

Guevara, Ernesto. 2006. *La Guerra de guerrillas*. Melbourne, Australia: Ocean Sur.

Juventud Rebelde. 2016. "Los jóvenes en la vanguardia defensiva." November 18. Accessed September 2, 2020. http://www.juventudrebelde.cu/cuba/2016-11-17/los-jovenes-en-la-vanguardia-defensiva.

Klepak, Hal. 2012. *Raúl Castro and Cuba: A Military Story*. Basingstoke, England: Palgrave Macmillan.

La Rosa, Lesmes. 1983a. “Bastión ’83: Así Será la Guerra de Este Pueblo.” *Verde Olivo*. June 30, 24 (26): 4–13.

La Rosa, Lesmes. 1983b. “Ensayo de Victoria.” *Verde Olivo*. November 10, 24 (45): 30–33.

Lezcano Lavandera, Yohana. 2013. “Ejército Central: ¡Decididos a Vencer!” *Bohemia*. November 29, 105 (24): 36–38.

Martín Blandino, Jorge. 2009. “La organización alcanzada inspira respeto.” *Granma*. November 28. Accessed September 2, 2020. http://www.granma.cu/granmad/secciones/bastion-2009/art16.html.

Mazarr, Michael J. 1990. “The Cuban Security Apparatus.” In *Cuba: The International Dimension*, edited by Georges A. Fauriol and Eva Loser, 257–292. New Brunswick, NJ: Transaction.

Parlamento Cubano. 2020. “Ley de la Defensa Nacional.” Accessed September 18, 2020. http://www.parlamentocubano.gob.cu/index.php/documento/ley-de-la-defensa-nacional/.

Pasanaut Suárez, Ramón. 2014. “Reseña de los Ejercicios Estratégicos Bastión.” *Cuba Defensa*. Accessed September 18, 2020. http://www.cubadefensa.cu/?q=ejercicio-bastion.

PCC. 2016. “Informe Central al 7mo. Congreso del Partido Comunista de Cuba, presentado por el Primer Secretario del Comité Central, General de Ejército Raúl Castro Ruz, La Habana, 16 de Abril de 2016, Año 58 de la Revolución.” April 2016. Accessed September 2, 2020. https://www.pcc.cu/sites/default/files/congreso/pdf/20180426/informe_central_vii_congreso_pcc.pdf.

Radio Cadena Agramonte. 2013. “Aviso a la población centra quehacer en Día de la Defensa en Cuba.” Accessed September 2, 2020. http://www.cadenagramonte.cu/articulos/ver/37990:aviso-a-la-poblacion-centra-quehacer-en-dia-de-la-defensa-en-cuba.

Rodríguez Rodríguez, Elvis R. 2016. “Guerra de todo el pueblo.” *Verde Olivo*. December 1. Accessed September 18, 2020. http://www.cubadefensa.cu/?q=node/4051.

Vanguardia. 2013. “Villa Clara, un bastión expugnable.” November 23. http://www.vanguardia.cu/villa-clara/217-villa-clara-un-bastion-inexpugnable.

Chapter 6

Meteoro

The Impact of Education on Disaster Preparedness and Risk Reduction

Emily J. Kirk

Education has been key to the revolutionary project. While literacy rates were arguably among the highest in Latin America before the Revolution, at roughly 76 percent, education, like health care, was largely centered in urban areas and reserved for the (often foreign) élite (Pérez 2008). Despite the deceptively high literacy rate, education was low and the shortage of teachers, schools, and educational opportunities across the island was noticeable, especially in rural and marginalized areas. As such, the improvement of education in terms of quality, accessibility, and equity became a central pillar of the Revolution.

Fidel Castro noted the importance of education within the revolutionary project well before its inception in 1959. In his now-famous "History Will Absolve Me" defense speech of 1953, he expressed his views on Cuba's education system and the needed improvements. Education, he noted, was one of the six problems the Revolution "would take immediate steps to solve" (Castro 1953). He explained:

> Where the peasant doesn't own the land, what need is there for agricultural schools? Where there is no industry, what need is there for technical or vocational schools? Everything follows the same absurd logic; if we don't have one thing we can't have the other. In any small European country there are more than 200 technological and vocational schools; in Cuba only six such schools exist, and their graduates have no jobs for their skills. The little rural schoolhouses are attended by a mere half of the school age children—barefooted, half-naked and undernourished—and frequently the teacher must buy necessary

school materials from his own salary. Is this the way to make a nation great? (Castro 1953)

Indeed, the revolutionary leadership made it clear that educational reform was key to the success of Cuba and its population well into the future.

This emphasis on education has continued throughout the decades, as the leadership sought immediate and sweeping changes. While solutions were often imperfect, the Revolution did ingrain in the Cuban population the importance of education. As a result, contemporary Cuba has one of the most educated populations in the world, distinguishing itself in particular among global south nations (Gasperini 2008; Smith 2018).

Cuba's notable education practices have maintained a crucial role within the island's disaster preparedness and risk reduction strategies. The country's educated (literate and well-informed) population, as well as the formal and informal education[1] dedicated to climate change and natural hazards, has inculcated in the population a profound understanding of the importance of disaster management and climate change adaptation. This has consistently been a dedicated focus of the revolutionary project since the early 1960s, building a society that, through education, has learned to prepare, act, and adapt.

A key example of the impact of education on Cuba's disaster preparedness strategies is known as *Meteoro*, a national simulation exercise. Beginning in 1986 and held every consecutive year, *Meteoro* is a two-day event in which the island's population is required to practice a wide variety of actions and drills related to natural hazards or health crisis, as well as prepare the island's infrastructure for the upcoming hurricane season. This includes practicing evacuation plans, rescue scenarios, ensuring emergency provisions, and clearing power lines from potential damage. As a result of this annual event, emphasizing education, participation, and solidarity, Cuba and its population maintain vigilance and are consistently prepared for major meteorological events, among other hazards. This chapter will briefly trace the evolution of education in post-1959 Cuba as well the development and significance of *Meteoro*. Education, especially as exemplified by *Meteoro* has, without doubt, been a decisive factor of the island's ability to protect life and property in times of disaster.

EDUCATION IN CUBA

Education has been an aspect of paramount importance for international development, particularly as a means of improving agency and wellbeing. For post-1959 Cuba, education has played a significant role in the island's development, as the country sought to dramatically improve quality of life through education.

For example, while the private education system was well established in pre-revolutionary Cuba, and despite well-trained teachers and well-built schools, the central problem was their location and access to schools. Although many of these schools were highly regarded, they were almost exclusively centered in urban areas and attended by the élite (Gasperini 2008; Leiner 1985). To put this in context, and as explored in some depth in Collins' chapter within this volume, in 1958, almost half of all children aged 6–14 had received no formal education. Moreover, only a quarter of those aged 15 years and older had ever attended school and three-quarters had not completed primary education (Pérez 2008, 273). As Rolland (1971, 385) described, "the Cuban school system provided strikingly unequal educational opportunities to students according to their socioeconomic status and place of residence. The system . . . fostered aspirations that were simply unrealistic for the vast majority of Cuban children and [was] largely dysfunctional for national development." However, it did not take long for this to change under the revolutionary leadership.

The year 1961 became the "Year of Education" as the population was mobilized in what was referred to as the "Literacy Campaign." The aim was to use the most educated among the population, including teenagers, to teach basic literacy to the population, with a particular focus on the most rural and marginalized areas (Smith 2018). The following year, the government reported that the adult literacy rate had increased to 96 percent, which was "the highest in Latin America and one of the highest in the world" (Pérez 2008, 273). After the initial push for literacy, campaigns and innovative strategies to improve education became standard, and free universal education was introduced. Schools too were built throughout the island, and teachers were trained to staff them. Paradoxically, within a period of a few years, education became a source of national pride, and slogans such as "the path up from underdevelopment is education" were visible throughout the country.

Within a decade of the "Year of Education," Cuba had effectively achieved the universalization of primary education, with 96.4 percent of children between 6 and 12 in formal education (Ferrán 1976). As Smith (2018) explained:

> This aim accomplished, attention turned to ensuring that students could continue their education at secondary level; the Battle for the Sixth Grade was superseded by the Battle for the Ninth, which had become the new minimum requirement for professional "intermediate level" roles. The belief in the importance of education as a means of development underpinned these ongoing and evolving developments. (Fitzgerald 1994)

The First Congress on Education and Culture occurred in April 1971.[2] The Congress was held in Havana, with 1,800 delegates from across the country. It was the product of some 116,00 education workers participating in 2,599 sessions at the municipal, regional, and provincial levels to debate the main areas regarding education (Leiner 1994, 35; UNESCO 1971, 5–6). The results were significant, as more than 3,000 recommendations were made by educational workers from across the island (Ferrán 1976). The result of the Congress was the nation-wide understanding that, while education had dramatically improved in the preceding decade, significant improvements were needed. A study on these recommended improvements was published in 1976, entitled the "Plan to Perfect and Develop the National Education System" (*El Plan de Perfeccionamiento y Desarrollo del Sistema Nacional de Educación de Cuba*). It outlined a five-year policy plan to reform the education system.

With the Plan to Perfect and Develop the National Education System in place, changes and improvements to the system came quickly. Of particular note, and meeting the demand of increased student enrolment, Basic Rural Secondary Schools (Escuelas Secundarias Básicas en el Campo, or ESBEC) were built in the countryside (Smith 2018). In essence, they were large boarding schools in rural areas, where students were expected to study and work in agricultural production (such as on local farms). In April 1972, Fidel Castro promised new schools offering places for 20,000 students that year, with an additional 80,000 in 1973 and 100,000 in 1974 (Castro Ruz 1972). The deficit of teachers to educate the increase in students also presented a problem. For example, in 1972, it was estimated that 18,548 new junior high teachers would be required over the succeeding four-year period, which was significantly more than those scheduled to complete their teacher training within that time frame (Ibid.). As an answer to the question of sufficient numbers of teachers, educated teenagers (as young as 15) were employed to lead classes. While in many ways these were flawed (and at times heavily criticized) solutions, the ESBECs did continue the Revolution's aim of universal and free education.

Although the island was faced with considerable challenges, mainly economic issues as a result of the U.S. blockade, collapse of the Soviet Union, and subsequent Special Period, education continued to be a key tenet and focus of the revolutionary project. For example, by the end of the 1980s, junior high school enrolment was 94 percent and high school or the vocational equivalent was 86 percent (Smith 2018). There were also significant increases in university places, which swelled from 35,137 during 1970–1971 to 205,000 in 1980–1981 (Fitzgerald 1994); university enrolment was seventy-seven times greater than the 1958 enrolment numbers (Smith 2018). By 1998,

94 percent of teachers held university degrees in education (Gasperini 2000). The understanding of education as a means of national development remained at the forefront of government policies, and the new millennium was met with a renewed commitment to universal education.

It is also instructive to place Cuba's educational achievements in terms of meeting the UN's Millennium Development Goals (MDGs) and subsequent Sustainable Development Goals (SDGs). In terms of MDGs, by the time the goals were established in 2000, the targets were already met in Cuba. For example, the MDG for education noted that "by 2015, children everywhere, boys and girls alike, will be able to complete a full course of primary schooling." However, in Cuba, by 2000, the primary completion rate was 96 percent and education was provided at no cost to students, and geographic distinctions became irrelevant (World Bank 2021a).

The SDGs, instituted in 2015 with the aim of being accomplished by 2030, were more comprehensive, building on the MDGs. Despite the broadening of targets associated with education, they continued to be met consistently in Cuba. Target 4.1, for example, sought to ensure by 2030 that "all girls and boys complete free, equitable and quality primary and secondary education leading to relevant and effective outcomes" (UN 2015). Within the Cuban system, by 2014 (before the SDGs were instituted), the primary completion rate was a steady 96 percent, while the global average was 90 percent; enrolment for secondary schools reached 100 percent, while the global average was 76 percent (World Bank 2021b). Literacy rates are also revealing, as Target 4.6 sought "literacy for all" by 2030. Although the global average in 2015 was 85.3 percent, Cuban rates already met, and maintained, its literacy rate of 100 percent (World Bank 2021c). A significant component of this achievement was the ability to have full control over the public education system rather than responding to directives from donor countries, the IMF, or World Bank.[3] This translated to the island's education system's ability to be coordinated and fully nationalized, evolving to respond to the needs of the domestic population.

Education in Cuba has long been considered laudable and exemplary, particularly for a middle-income country. This has largely been the result of the government's consistent investment in education, which spends between 10 and 11 percent of the island's annual GDP on education—almost double the amount recommended by the UN. Independent from the Western, global north development industry, Cuba's education system was able to adapt to the evolving needs of, and challenges faced by, its population. It is this distinguished history of education that laid the groundwork for the island's successful track record of disaster preparation and risk reduction. *Meteoro* perhaps exemplified this best.

METEORO

Meteoro was initially inspired by the Civil Defense's (Defensa Civil or DC) guidelines regarding the protection of the Revolution. In addition to formal education, *Meteoro* is deeply rooted in the DC and its ideology of protecting the revolutionary project from internal and external threats. As noted by the DC, "the National Civil Defense and specialists from other institutions consult and assess past historical documents and methodologies used to prepare the population and governing bodies" (Defensa Civil 2014). This has evolved to largely focus on hazards, including hurricanes, tropical storms, floods, fires, as well as health threats, such as infectious diseases.

Hurricane Kate hit Cuba in November 1985, causing significant damage and chaos. As a result, flaws in the island's preparation, particularly related to the mobilization of the population, were highlighted. While the DC was previously charged with responding to threats of natural hazards at all levels, including preparation for such natural disasters, the shortcomings evidenced by Hurricane Kate highlighted the need to expand preparation strategies—not simply against the threat of war but for the protection of the population, infrastructure, and economy against the increasing threat of disasters. As such, *Meteoro* emerged as one of the DC's strategies for disaster preparedness in 1986, and has been held ever since.

Meteoro can best be understood as a national preparation exercise. Originally, it was mainly dedicated to preparing the population ahead of hurricane season. However, in the 1990s, it expanded its objectives to include other natural hazards as well as technological and health risks. Since 1998, *Meteoro* has typically been carried out during a weekend in late May, with the first day characterized by practicing possible situations, while the second day is spent securing infrastructure (Sandoval Guzmán 2014). According to the DC, it was important for the population to actively practice drills to see how plans worked in reality rather than just in theory (Defense Civil 2014). The DC increasingly focused on preventive measures (including preparation and mitigation strategies) rather than responsive measures. As natural hazard governance and risk reduction specialists Castellanos Abella and Wisner (2019, 12) explained:

> Initially the exercise was designed to better prepare the population for hurricane season (June–November), but gradually it started to include other disaster types at all disaster management levels with heavy involvement of the local population. *Meteoro* provides a test of the civil defense system. Also during this period, sub-national jurisdictions and local authorities started to have a broader view of the hazards they faced, and they began to consider many diverse scenarios including dam failure, chemical contamination, epidemics, and so forth.

In essence, *Meteoro* makes actively practicing for a given disaster possible. As described by Thompson and Gaviria (2004, 33), "Once a year, at the end of May, the whole country, in their respective ministries, schools, hospitals, and factories participates in a two-day training exercise in risk reduction." On the first of the two-day exercise, the DC determines what a given disaster may be and where (affecting a combination of evacuation plans, population and livestock safety, including complex rescue missions, as well as economic challenges or infrastructure damage). Other activities include community risk mapping, assessing and updating data of vulnerable populations and their needs (such as elderly or disabled community members), evaluating water safety measures, and determining the safest routes to shelters, among others. It then becomes the responsibility of the community and/or region to mobilize, work collaboratively, and actively demonstrate how the problems will be solved.

In 2003, as *Meteoro* was continuing to evolve and adapt to meet the contemporary needs of the population and increasing weather phenomena, Civil Defense representative in Cienfuegos, José Castro, explained the event as:

> In this exercise we try out everything. We give people simulation exercises to do. . . . For example, with the electric company here in Cienfuegos, the head of the Civil Defense will give the electrical company's director a situation: "So many lines are down, so many generators are affected, and here is your main problem. . . . " And he will be asked, "How will you confront this situation?" He has to respond based on his emergency plan outline. On the first day, everyone engages in these exercises, responding to hypothetical situations. On the second day, they carry out all the physical preparation measures in the community. (Thompson and Gaviria 2004, 33)

One of the main exercises, which has been routinely practiced, includes the first three (of four) stages of the hurricane preparedness and response strategy. These include "Information" (which occurs approximately 72 hours before landfall), "Alert" (implemented 48 hours in advance of an expected hazard), and "Alarm" (activated 24 hours in advance). The fourth is characterized as "Response" (which occurs when the government and DC note the danger has passed) (Kirk 2017). Specific actions for each stage are designated as well as practiced over the course of *Meteoro*.

In addition to the active participation at the community, municipal, provincial, and national levels, *Meteoro* includes detailed assessments of the exercises that were undertaken. Members of the DC as well as specific commissions review and analyze if the exercises were successful, ultimately determining areas of needed improvement. This occurs with the view to annually improve planning and preparation so the country can continue updating risk reduction strategies and increase their efficacy; drills are routinely updated based on the latest results of risk reduction studies.

Further to post-*Meteoro* assessments, large-scale vulnerability studies are common throughout the country and are also used to update disaster risk reduction strategies. Indeed, a given vulnerability study includes a wide variety of specialists, working collaboratively to determine areas of concern and best practices. For example, a multi-disciplinary provincial study conducted over a two-year period on vulnerability to rain, wind, and sea flooding in Holguín included the participation of three sociologists, a geographer, a mathematician, two civil engineers, and healthcare and housing technicians. The study identified which streets should be evacuated first, depending on the hazard. These results were subsequently included in the updated *Meteoro* plans for the area and practiced by the communities (Sandoval Guzmán 2014).

Following the first day of practiced exercises, the second day of *Meteoro* largely consists of widespread community mobilization to prepare and protect infrastructure. Communities are responsible for their own municipality (as each often requires different preparations depending on geography and potential hazards). Community members can be seen, for example, pruning tree branches to protect power lines, securing objects and infrastructure (such as placing tape across windows to minimize the potential spread of broken glass), and cleaning debris or garbage from streets and sidewalks (this serves to limit draining blockages during heavy rains and floods, as well as minimizing the risk of water contamination leading to communicable diseases).

Of particular note, *Meteoro* is not a repetitive process of the same drills. It has continued to evolve since its inception, adapting to climate change, updated meteorological data, and contemporary needs. For example, in 2010, *Meteoro* was extended to include two parts, including potential new hazards. The first part, occurring May 21–23, followed the general organization structure of drills and preparation for hurricanes, as well as a variety of health emergencies. However, the second stage, occurring June 25–26, focused on how to respond to the extreme natural hazards that were increasingly occurring globally. It was determined by the DC and government that Cuba needed to be prepared for other "extreme situations," which could also impact Cuba in the future. Information and preparation strategies centered specifically on large-scale earthquakes and tsunamis (López Pino 2010).

The following year, as well as between 2011 and 2014, the aim of *Meteoro* was developed again to meet the challenges of that period. Specific "objectives" and "practical activities" were designated by the DC to be implemented by the given governing bodies at the community, municipal, provincial, and national levels (Defensa Civil 2014). These included:

Objectives

- Analysis of the major guiding documents for the reduction of disasters, with emphasis on an objective analysis, while ensuring the measures included in the plans for each of the respective phases, starting from the risk analysis undertaken on the type of danger involved.
- Evaluation of the structure, composition, and work conditions required for the functioning of the Control Point of the Defense Council which would allow the effective and efficient management of actions to both respond and recover in disaster situations.
- Analysis of the information system to guarantee a supply of objective analysis in the decision-making process of the President of the Defense Council as well as to check and exchange information between different levels of the organization.

Practical Activities

- Increasing the level of preparation of the population in disaster reduction measures within their community, corresponding to the appropriate risks in each region.
- Increasing the state of preparedness of those forces that participate in response and recovery measures in disaster situations, including the cooperation of the Revolutionary Armed Forces (Fuerzas Armadas Revolucionarias, or FAR) and the Ministry of the Interior (Ministerio del Interior, or MININT), according to the disaster reduction plans.
- Carrying out mitigation activities in terms of vulnerabilities identified during the risk analysis process.
- Checking the effectiveness and assurance level of the ways and forms to both inform and guide different categories of the population in disaster situations.

The focus was ardently placed on prevention and the minimization of losses of human lives, materials, and infrastructure. Mario Enrique García Salas, DC leader within the Provincial Defense Council of Havana, stated in 2012, "We are carrying out a series of preventive actions with the aim of reducing vulnerabilities and dangers that impact the population, before any events occur" (Morales Agüero and Váquez Muñoz 2012).

Although directed by the DC, it is important to note that each major sector also actively participates in *Meteoro*. It should be emphasized that the pillars of the Revolution—the healthcare and education systems—play an active role in terms of local and national disaster readiness. Specifically, healthcare professionals and educators are responsible for schools and hospitals, and in addition function as community leaders in terms of prevention as well in times of disasters.

Education is broadly important for a knowledgeable and literate population including a wide array of trained professionals. However, in terms of disaster preparedness, the formal education system has also been key to teaching new generations. Beginning in third grade, the national curriculum includes lessons on the environment, climate change, disaster risk reduction, and disaster response. These include, for example, disaster risk mapping, identifying evacuation routes depending on the hazard, diverse agricultural practices, energy conservation, renewable energy sources, and the complexities of climate change at the global, national, provincial, and municipal levels (Santos Pérez et al. 2018).

Significant publications on these, among other, topics are commonplace throughout the island, particularly directed at children and young people. Largely organized, published, and distributed by the DC, they note disaster preparedness and risk reduction strategies that are practiced within *Meteoro*. For example, the *Family Guide for Protection Before Tropical Hurricanes* (*Guía familar para la protección ante ciclones tropicales*) (Defensa Civil n.d.) is directed at children and explains in detail the preparation and safety measures before, during, and after a hurricane. Included in the guide are possible situations depending on one's location and the variable risks, such as if the hurricanes occur in a rural or urban area.

Disaster management is also included throughout secondary school, post-secondary, and graduate education. Disaster mitigation and climate change adaptation are included in a broad range of professional training, from tourism to agriculture. For example, and as discussed in Mesa Ridel's chapter in this volume, Cuba's health sector includes extensive and specialized training regarding natural disasters and climate change. Disaster Medicine (Medicina de Desastres) has been actively included in undergraduate training since 2004 and was later expanded to graduate-level training, under the broader designation of Health and Disasters (Salud y Desastres). Undoubtedly, Cuba's formal education system, ranging from coursework to practical application of disaster mitigation strategies, has been crucial to the island's success in disaster preparation and risk reduction.

CONCLUSION

Ramón Pardo Guerra, the head of the National Civil Defense, affirmed "*Meteoro* is undoubtedly a strength of Cuba. Thanks to the [revolutionary] system we have implemented and consolidated for more than 50 years, we have been able to face any situation" (Cáceres 2014). However, it is unlikely that *Meteoro* would either exist or be successful without the strong basis of education that developed throughout post-1959 Cuba; it illuminates the value of the country's dedication to education. The development of comprehensive and equitable education nationwide laid the groundwork for both educated and engaged professionals, as well as the population with a firm grasp on the importance of preparation and mobilization to reduce risks in times of disasters.

Discussed in classrooms at all levels, and practiced annually since 1986, the weekend-long drill has become a key component of Cuba's approach to disaster management and risk reduction. Ultimately, *Meteoro* illustrates the importance of education—both inside and outside of the classroom. While perhaps not directly transplantable as a strategy elsewhere, there is certainly a lot that other nations can learn from Cuba's experience, the role of education, as well as the evolution and impact of *Meteoro*. Surely, while adverse meteorological events are increasing in frequency and ferocity, outlining, updating, and practicing emergency measures could only help to save lives, protect infrastructure, and mitigate disasters elsewhere.

NOTES

1. In this context, formal education can be understood as programs and curricula organized and implemented by the Cuban state. This includes standardized primary, secondary, and higher education, as well as directives and exercises organized by state representatives or entities, such as *Meteoro*. Informal education refers to education that occurs outside of formal educative practices, such as information or direction received from family members and neighbors or non-state-run media (such as radio and visual media).
2. While known in many ways as being famous for ushering in the repressive cultural policy of the *quinquenio gris* (five grey years), in the context of government-led formal education, the Congress was an important milestone.
3. As a result of the US embargo, Cuba was not permitted to engage with main international trade and loan organizations. This, in turn, meant that Cuba developed independently from global north development strategies attached to loan agreements, such as widespread privatization and adherence to Structural Adjustment Programs (SAPs).

BIBLIOGRAPHY

Cáceres, Patricia. 2014 "Meteoro 2014: una Fortaleza de Cuba." *Juventud Rebelde*. May 17. Accessed November 14, 2020. http://www.juventudrebelde.cu/cuba/2014-05-17/meteoro-2014-una-fortaleza-de-cuba.

Castellanos Abella, Enrique A., and Benjamin Wisner. 2019. "Natural Hazards Governance in Cuba." *Oxford Research Encyclopedia of Natural Hazard Science*. Accessed February 15, 2020. https://www.researchgate.net/profile/Enrique_Caste

llanos2/publication/330635389_Natural_Hazards_Governance_in_Cuba/links/5c4b321fa6fdccd6b5c85750/Natural-Hazards-Governance-in-Cuba.pdf.
Defensa Civil. n.d. *Guía familar para al protección ante ciclones tropicales.*
Defensa Civil. 2014. "Meteoro: Acciones en caso de desastres." Accessed December 15, 2020. http://www.cubadefensa.cu/?q=meteoro-acciones.
Defensa Civil. 2017. *La Defensa Civil Cubana.* Havana: Casa Editorial Verde Olivo.
Ferrán, Héctor, editor. 1976. *El Plan de Perfeccionamineto y Desarrollo del Sistema Nacional de Educación de Cuba.* Havana: Ministerio de Educación.
Fitzgerald, Frank T. 1994. *The Cuban Revolution in Crisis: From Managing Socialism to Managing Survival.* New York: Monthly Press.
Castro Ruz, Fidel. 1953. "History Will Absolve Me." Accessed January 17, 2020. https://www.marxists.org/history/cuba/archive/castro/1953/10/16.htm.
Castro Ruz, Fidel. 1972. "Discurso pronunciado por el Comandante Fidel Castro Ruz, primer secretario del comité central del Partido Comunista de Cuba y primer ministro del gobierno revolucionario, en la clausura del II congreso de la Unión de Jóvenes Comunistas, efectuada en el teatro de la CTC-Revolucionaria, el 4 de abril de 1972." Accessed November 17, 2020. http://www.cuba.cu/gobierno/discursos/1972/ esp/f040472e.html on 19.12.2017.
Gasperini, Lavinia. 2000. "The Cuban Education System: Lessons and Dilemmas." *Country Studies: Education Reform and Management Publication Series* 1 (5).
Gasperini, Lavinia. 2008. "The Cuban Education System: Lessons and Dilemmas." In *Reinventing the Revolution: A Contemporary Cuba Reader*, edited by Philip Brenner, Marguerite Rose Jiménez, John M. Kirk and William LeoGrande, 299–301. Lanham, MD: Roman and Littlefield.
Kirk, Emily. 2017. "Dealing with the Perfect Storm: Cuban Disaster Management." *Studies in Political Economy* 98 (1): 93–103.
Leiner, Marvin. 1985. "Cuba's Schools: 25 Years Later." In *Twenty-Five Years of Revolution 1959–1984*, edited by Sandor Halebsky and John M. Kirk, 27–44. New York: Praeger.
López Pino, Maite. 2010. "Meteoro 2010: Sismos y Maremotos en Cuba: ¿Qué hacer?" *Cubahora.* May 21. Accessed November 14, 2020. https://www.cubaheadlines.com/2010/05/21/21716/sismos_y_maremotos_en_cuba_%C2%BFque_hacer.html.
Morales Agüero, Juan, and Luis Raúl Váquez Muñoz. 2012. "Un ejercicio estratégico para el país." *Juventud Rebelde.* May 19. Accessed November 14, 2020. http://www.juventudrebelde.cu/cuba/2012-05-19/un-ejercicio-estrategico-para-el-pais.
Paulston, Rolland G. 1971. "Education." In *Revolutionary Change in Cuba*, edited by Carmelo Mesa-Lago, 375–398. Pittsburgh: University of Pittsburgh Press.
Pérez, Louis A. 2008. *Cuba Between Reform and Revolution*, 3rd edition. Oxford: Oxford University Press.
Sandoval Guzmán, Javier. 2014. "Popular Power and Environmental Governance: The Cuban Approach to Natural Hazards and Disaster Risk Reduction." MA diss., Norwegian University of Life Sciences.
Santos Pérez, Santiago, Alfonso Núñez Leguen, Marbelis Rodríguez Azahare, Virginia Huergo Silveira, Milagros Santa Cruz Domínguez, Berta Lidia Castro, and

Mayra Díaz García. 2018. *Buenas prácticas cubanas para la protección de niños y adolescentes en situaciones de desastres*. Havana: Sello Editor de la Educación Cubana.

Smith, Rosi. 2018. "The 'Three Ps' (Perfective, Professionalization, and Pragmatism) and their Limitations for Understanding Cuban Education in the 1970s." In *Cuba's Forgotten Decade*, edited by Emily J. Kirk, Anna Clayfield, and Isabel Story. Lanham, MD: Lexington Books.

Thompson, Martha, and Izaskun Gaviria. 2004. *Weathering the Storm: Lessons in Risk Reduction from Cuba.* Boston, MA: Oxfam America.

UN. 2015. "Sustainable Development Goal 4." Accessed January 21, 2021. https://sustainabledevelopment.un.org/sdg4.

UNESCO. 1971. "XXXIII Conferencia Internacional de Educación OIE-UNESCO, 15–23 Septiembre." Informe de Cuba. Organización de Educación (UN. Doc. OIE/Q/70/3). Accessed October 20, 2020. http://www.ibe.unesco.org/National_Reports/Cuba/nr_mf_cu_1971_sef.pdf.

World Bank. 2021a. "Primary Completion Rate." Accessed January 21, 2021. http://data.worldbank.org/indicator/SE.PRM.CMPT.ZS.

World Bank. 2021b. "Gross Enrolment Ratio, Secondary." Accessed January 21, 2021. http://data.worldbank.org/indicator/SE.SEC.ENRR.

World Bank. 2021c "Adult Literacy Rate." Accessed January 21, 2021. http://data.worldbank.org/indicator/SE.ADT.LITR.ZS.

Chapter 7

Postgraduate Education Concerning Natural Disasters and Climate Change in the Cuban Health Sector

Guillermo Mesa Ridel
Translated by Anna Clayfield

Natural disasters constitute one of the principal threats to stability and development in society. Exposure to these phenomena has resulted in different disaster situations across the world, which coincide with, and are exacerbated by, other risk scenarios, such as globalization, international migration, displacement, mass-casualty incidents, and armed, ethnic, economic, political, or social conflicts. All of these scenarios, together with other, equally complex phenomena—those relating to people, food, and the environment, for example—affect human life by impacting individual and population health and health institutions themselves (Mesa Ridel et al. 2018). The latter are already being adversely affected by climate change on a global scale, given their vulnerability to climate-related stress and the eroding capacity of health systems to manage the changing patterns of effects on health (World Health Organization 2018).

Many disasters occur in large part as a result of a lack of information and timely and adequate communication among people and society. In light of contemporary advances in science and technology, a change is occurring in the ways in which we access information and knowledge. This change has itself created a gap between information consulted online and that which circulates in the world, between knowledge acquired through formal education and that which is needed to succeed professionally, as well as between knowledge and know-how more generally. With this gap in mind, information and knowledge management is of unique value to current and future generations (Fernández Valdés and Zayas Mujica 2016).

Efforts to improve disaster-related training of human resources for health have been justified in the Americas region by the marked increase in the impact of adverse events and the accompanying rise in human, material, economic, and environmental costs. As such, disaster programs have been introduced as part of professional development, although they are limited to pre-hospital care, as well as medical and surgical specializations; they essentially center on the care provided by hospital emergency departments and scarcely focus on the intersectoral work required to manage disaster risk reduction. Moreover, while many programs have been developed on-site, these are mostly concentrated in large cities and create barriers to accessibility (Guerrero Africani 2016; Paniagua 1995).

This chapter aims to summarize the Cuban health sector's experience of skills development in human resources during the period 1996–2004; first launched as Disaster Medicine (Medicina de Desastres) as part of undergraduate training, it later extended to post-graduate education under the new banner of Health and Disasters (Salud y Desastres). The chapter traces the evolution of the Latin American Center for Disaster Medicine (Centro Latinoamericano de Medicina del Desastre, or CLAMED) in terms of disaster risk reduction, illuminating the key challenges facing post-graduate education from an institutional perspective and with a view to the new global threats posed by climate change. In so doing, it reveals how CLAMED's initiatives comply with the legal and regulatory framework that oversees the planning and organization of Cuba's efforts in these respects.

KNOWLEDGE: THE FOUNDATION OF DISASTER RISK REDUCTION

Owing to its geographical location in the western Caribbean Sea, Cuba is exposed to various natural hazards, in addition to those of a health and technological nature. Every year, it is affected by hydrometeorological phenomena, particularly tropical hurricanes. The country's hurricane season spans six months, from June 1 to November 30, during which time the Cuban territory has, over the years, seen damage to housing, agricultural, and industrial infrastructure. Additionally, some communities in isolated, coastal, and mountainous areas are left *incommunicado* for several days until normal conditions are restored. In the last thirty years, however, the number of human casualties has been reduced thanks to early warning and population protection systems (Ramos Guadalupe 2009).

Civilian protection measures in Cuba before 1959 suffered from a narrowly conceived approach to health. Actions taken were essentially limited to medical care offered by *Casas de Socorro* (primary care centers), under the aegis of Cuba's former Ministry of Health, Ministerio de Salubridad, and the services of the Fire Department (Cuerpo de Bomberos). Both bodies lacked effective government support and, in general, no attention was devoted to training professionals in health institutions to reduce the impact of disasters.

After 1959, the Cuban leadership adopted the first disaster risk reduction measures. In 1966, the state-level System of Civil Defense (Defensa Civil, or DC) Measures was instituted in law, with the principal function of safeguarding the population as well as social and economic achievements from hazards, natural or otherwise. Presidents of provincial and municipal governments were delegated responsibility for directing the DC in their respective jurisdictions. Specifically, this involved implementing measures that were coordinated and executed by state bodies, economic entities, and social institutions, with the full participation of the entire population. In the event of, or just prior to, a disaster or similar situation, the aforementioned law also regulated when a state of emergency could be declared and brought to an end, as well its intermediate effects (Asamblea Nacional del Poder Popular 1966, 1974; Castellanos 2012).

Over the years, the DC system has been improved based on lessons learned and lived experiences. It has developed a vast legal framework that encompasses laws, decree laws, decrees, and ministerial resolutions. This provides a solid institutional base for its implementation and allows for compliance monitoring. Decree Law No. 170 on the System of measures of Civil Defense of May 1997 defined the features related to disaster risk reduction in Cuba and established its specific body of law, in addition to the structural and educational measures that have since positively impacted social, economic, and security indicators for the population (Consejo de Estado 1997; República de Cuba 1999).

One of the first actions undertaken by the DC after 1963 was the conception and development of a system of hydraulic works in areas at risk of flooding, designed to attenuate and control the floodwater produced by periods of heavy rainfall. This system, known as "Voluntad Hidráulica" (Hydraulic Drive), allows for continuous reporting on rainfall patterns, as well as data related to the functioning of these hydraulic works. As a result of its efficient operation during extreme circumstances, the system guarantees the management of risk factors associated with floods and enables the effective protection of the population and economic resources.

In 1987, Resolution 169 of the 42nd session of the General Assembly of the United Nations urged the global community to pay special attention to international cooperation with regard to disaster reduction. With the announcement of the International Decade for Natural Disaster Reduction 1990–1999, a framework of action was approved to improve national capacities, particularly in developing countries. In December 1999, the general assembly adopted the International Strategy for Disaster Reduction (ISDR). Consequently, disaster risk reduction went from being a narrowly perceived technical discipline to a broad global movement focused on sustainable development and recognized by governments and organizations as a means of increasing community resilience.

Within this framework, diverse international institutions achieved progress in training health professionals and the general community. Training activities, including specialist technical training in conflicts and humanitarian emergencies, were also introduced. Additionally, Disaster Information Centers were established based on collaborative efforts around building a culture of prevention, while advances were made in disaster-focused Prehospital Emergency Medicine and Public Health (Sánchez 2000; WADEM n.d.).

In the 1990s, Cuba created the Veterinary Center for Prevention in Cases of Disaster (Centro Veterinario para la Prevención de Desastres), with a view to revising veterinary training programs at every level and in all specialist areas. It also aimed to promote courses and events, in addition to encouraging research, the production of scientific literature and international collaboration through the creation of the Society for Veterinary Disaster Medicine (Sociedad de Medicina Veterinaria de Desastres) (Chávez Quintana 2010). For its part, the José Antonio Echeverría Higher Polytechnic Institute (Instituto Superior Politécnico José Antonio Echeverría) developed the Center for Disaster Prevention and Mitigation (Centro de Prevención y Mitigación de Desastres, or PREMIDES-CECAT) in the field of engineering.

These steps, together with the willingness of the Cuban people to offer assistance, in the spirit of solidarity, to any country in the world affected by disasters, constituted the precursors to the creation of CLAMED in 1996, under the guidance of the Ministry of Public Health (Ministerio de Salud Pública, or MINSAP). It was established at the request of the heads of National Civil Defense (Defensa Civil Nacional, or DCN), with the co-sponsorship of the Pan American Health Organization (PAHO), World Health Organization (WHO), and the Caribbean Medical Association (CMA).

CLAMED brought together scientific potential at the highest level of prevention, care, and research within Cuba's National Health System. It also acted as an important instrument for improving the coordination of actions against disasters and for a comprehensive strengthening of the scientific knowledge bases and prevention and mitigation methods in the field of human health. It took advantage of the National Health System's accumulated experience, resources, and institutions and adopted a multidisciplinary, intersectoral, and rational focus. Its incorporation of human resources training allowed it to gain influence over the preparation of the wider population regarding natural disasters (Mesa Ridel 2003).

CLAMED's founding president, Abelardo Ramírez Márquez, first vice minister of MINSAP, and representative to the National Commission for the International Decade for Natural Disaster Reduction, created the multidisciplinary teaching-focused Department of Disasters and Health for the education and training of health professionals. Its creation supported the health system's plans for disaster prevention and mitigation, with the wider objective of establishing a thematic network on health and disasters to disseminate up-to-date scientific information (Ibid.). In 1999, CLAMED's Center for Information and Documentation was inaugurated, on the back of an initial project funded by the European Commission Humanitarian Aid Department's Disaster Preparedness Programme in Latin America and the Caribbean (DIPECHO-LAC).

The basis for disaster reduction in Cuba is knowledge of hazards, vulnerabilities, and risks. Decision-making is underpinned by one key principle: protecting human life, the economy and the environment from the destructive effects of disasters and climate change. To this end, priority is given to knowledge regarding preparation, response, and recovery and to an understanding of measures that prevent and reduce vulnerabilities and risks and that will mitigate the impact of hazards in general. This knowledge contributes to a systematic process, which involves all of Cuban society to confront every type of hazard, to minimize impact and to achieve a speedy recovery at a national level in the face of any eventuality. In addition, human resources training and the planning put in place by the National Health System have also enabled the fulfillment of international cooperation commitments, without affecting the health service coverage of the Cuban population itself. Such cooperation has reinforced the need to rely on professionals from different specialist backgrounds who are prepared to work in countries affected by disasters (Marimón Torres and Martínez Cruz 2010).

Knowledge management has been facilitated by the desire of the Cuban state to attain a high level of education and of disaster reduction among the population, as well as by the culture of risk prevention that motivates organizations, institutions, and society as a whole. It also relies on an education policy—universally applicable across the health sector—which is preventive and curative, scientific, patriotic as well as internationalist, and has solid educational and instructive objectives. This approach enables optimal performance in prevention, preparedness, response, and recovery in the event of a disaster (Vela Valdés 2017).

HUMAN RESOURCES PREPAREDNESS FOR DISASTERS

Cuba's higher education system recognizes that many hazards are inevitable, particularly those of a natural origin, but that human action also makes it possible to eliminate risks, either totally or partially, by reducing vulnerabilities, increasing resilience, and improving the capacity to confront danger (Évora Capote 2012). Through the DC, students have at their disposal a preparation system with core components incorporated into all fields of study and with a multidisciplinary and transdisciplinary focus. Recent years have seen the implementation of a dynamic process to improve the system's overall quality, geographical accessibility, equality of opportunity, and exploitation of human resources and local materials, in conjunction with scientific research regarding solutions to the country's problems (Quesada Romero 2014).

Course curricula take into consideration the different stages of disaster risk reduction through practical activities, scientific conferences, workshops and exercises, and end-of-semester/degree final projects of considerable academic weighting. The initial induction process for new students arriving from other countries includes a code of conduct regarding disasters, in addition to epidemiological surveillance efforts in compliance with Cuba's own International Health Monitoring (Control Sanitario Internacional) (Sánchez Tarragó and Arriola Martínez 2014) system, as well as the WHO's International Health Regulations; the latter is aimed at preventing the spread of diseases and putting into practice control measures, social communication, as well as evaluation and readjustment where necessary (Gala et al. 2012). Teachers also receive information on relevant issues pertaining to the DC, with individual, systematic study representing the main means of improving and updating their skills (Oramas González et al. 2013).

Preparedness programs are also developed for managers, officials, and governing bodies, together with the participation of teachers and students, as part of *Meteoro*, the nationwide disaster simulation and preparation exercise. This annual event, aimed at enlisting recruits for hurricane season, involves the updating of plans, practical exercises and demonstrations, and sanitation and cleaning, among many other tasks, depending on the natural hazard in question (García Gómez 2016; López Nicles et al. 2018). Non-professional university workers also undergo a program tailored to the hazards associated with their specific roles, with particular emphasis on personal protection standards (Évora Capote 2012). All these elements have enabled the development of a Cuban pedagogical model, guided by the DC, which is evident in knowledge systems, teaching methods, the pedagogical and didactic training of the faculty, and the management of the teaching and learning process.

Regarding health care, MINSAP is responsible for the development of its own human capital when it comes to higher education. Course curricula are designed around the population's principal health problems, while the training process focuses on the integration of teaching, care and research, education in the workplace, and the interaction between the medical school and the community. The aim is to produce professional Cuban and foreign nationals who work both at a national level and as part of internationalist medical brigades (Vela Valdés et al. 2018).

Following the creation of a single, state-led, National Health System in 1960, training gradually extended to every province as part of a network of centers which, since 1976, has been under MINSAP's control and is methodologically regulated by the Ministry of Higher Education (Ministerio de Educación Superio, or MES). Of its many faculties, its most notable include those of medical sciences, dentistry, nursing, health technology, the Latin American School of Medicine (Escuela Latinoamericana de Medicina, or ELAM) and the National School of Public Health (Escuela Nacional de Salud Pública, or ENSAP). ENSAP is responsible for the training and development of human capital in public health at the post-graduate level, such as doctorates, specializations, masters, and other organizational forms of teaching (Vela Valdés et al. 2018).

MINSAP forms part of the structure of the DC system at all levels of the country's political-administrative division. Protected by legal standards, it is charged with the planning and organization of civil protection measures and is authorized to issue regulations regarding the surveillance, prevention, and control of epidemic diseases; to develop laws specific to territories and health institutions; and, through the Cuban Red Cross, to prepare the community for light search and rescue operations, first aid, and physical and psychological support for victims. In the event of a disaster, it becomes a component of the body of socioeconomic work required to implement decisions related to the protection of life and economic resources (Mesa Ridel et al. 2018).

The introduction of disaster-related content at the undergraduate level in Cuban medical schools began in the 1970s. The first stage was characterized by the teaching of the Epidemiology of Disasters and was limited to addressing basic issues concerning familiarization with different situations and the key medical requirements as part of a health-based approach, with a particular focus on outbreaks of infectious diseases (Bello Gutiérrez 2011).

The starting point for the teaching of disaster medicine came some two decades later in the year 1998, when the impact of two hydrometeorological events caused extensive damage in Central America and the Caribbean: Hurricanes Georges and Mitch. The Cuban leadership offered healthcare provision in the form of medical brigades to affected populations as part of its Comprehensive Health Program (Programa Integral de Salud, or PIS), which later included the training and development of human resources. It also offered opportunities for the mass training of doctors in Cuba following the creation of ELAM. By the end of that year, the teaching of disaster medicine had much improved, becoming a subject taught independently to the study of medicine, equipped with a body of knowledge and skills and with educative and instructive objectives (Bello Gutiérrez 2011; Marimón Torres and Martínez Cruz 2010).

By bringing continuity and sustainability to the PIS, ELAM constituted a core element of the country's solidarity initiatives. Through its development, the teaching of disaster medicine as an independent subject was consolidated. The course lasted 160 hours, taught over two sessions in the fourth and fifth years of a medical degree, and its original program gradually came to form part of the curriculum for all health-based university degrees across the country: medicine, dentistry, nursing, psychology, and health technology (Bello Rodríguez et al. 2011; Ferreiro Rodríguez 2015; Hernández Roque et al. 2014; Mejías Chao et al. 2015; Vidal Ledo et al. 2004).

Today's Cuban health professional is trained with knowledge of the hazards, vulnerabilities, and risks to the community and health institutions. They are formed under an effective leadership that interrelates the social determinants of health and risks, as well as implements effective preventative actions and oversees their comprehensive evaluation, while keeping the community for which they work at the center of everything they do (Teja Pérez and Mesa Ridel 2017).

CLAMED: A NEW FOCUS

Between the years 2004 and 2014, the Latin American Center for Disaster Medicine, CLAMED, slowly integrated the accumulated experience on disaster reduction into its practices. It became a systemic focus of human resources training, based on an evaluation of multi-hazards, a study of vulnerabilities, risk analysis, and a prioritization of the measures in the Disaster Risk Reduction Plan (Sauchay et al. 2016).

Alongside Disaster Medicine's key precepts, CLAMED included the identification of measures and actions, according to the stages of the disaster reduction cycle. On the basis of general risk analysis, it added specific actions to deal with emergencies in medical care institutions and the organization of cooperation, management, and communication. Each step takes into account all types of hazard (Mesa Ridel et al. 2018). CLAMED also played a fundamental role in the strengthening of local capacity, and the transfer of technology and collaborative projects, all with a multidisciplinary, intersectoral, and social participatory focus.

In just fifteen years of continuous efforts, a large number of professionals and specialists from highly regarded scientific institutions in the health sector were able to form an expert committee, which combined its work with that of specialists outside of the sector. This collaboration enabled the identification of key elements that helped to improve disaster reduction planning and organization in the national health system, and also contributed knowledge and positive results from scientific research, as well as cooperation in community work.

A comprehensive professional development strategy was then consolidated across the entire national health system, based around degree programs, courses, workshops, conferences, training sessions, and internships. The strategy was aimed at professionals and specialists linked to the practice of medicine and other branches of Medical and Health Sciences, in addition to other sectors related to health in general. The master's program in "Health and Disasters," established in 2007, produced national and foreign national graduates who contributed relevant results toward resolving problems and needs in their respective institutions (Ferreiro et al. 2014). CLAMED also provided training for Red Cross officials and directors, amateur radio operators (who play a key role in disseminating information in the event of an emergency), and managers, staff, and students of the Medical Sciences University of the Revolutionary Armed Forces (Universidad de Ciencias Médicas de las Fuerzas Armadas Revolucionarias). Since 2012, CLAMED has strengthened its offering of the annual international course "Health Emergencies in Large Populations (HELP)," which delivers training in the methods, practices, and principles of humanitarian aid, in conjunction with the International Committee of the Red Cross (ICRC) (Ferreiro Rodríguez et al. 2017).

Multiple educational workshops also took place in each of the country's provinces, while technical support was rolled out for regional groups that evaluated the vulnerability of health institutions and the professional development of practitioners and teachers in all of the medical science universities. At the same time, documentation was enriched with scientific literature, which helped to improve the quality of disaster risk reduction plans.

Training for Cuba's international medical brigades before undertaking their missions has also been systematic and ongoing. In 2005, training began for the Henry Reeve International Contingent of Doctors Specialized in Disaster Situations and Serious Epidemics (also known as the Henry Reeve Brigade). Since then, it has completed missions in more than twenty countries, including care for victims of the Ebola epidemic, which devastated Guinea-Conakry, Sierra Leone, and Liberia in 2014. This mission saw the contingent awarded the Dr. Lee Jong-wook Memorial Prize for Public Health based on the work and altruism of its members (World Health Organization 2018).

The contribution of a broad faculty of highly experienced and respected instructors from various disciplines has meant that the contents of the postgraduate curriculum in health and disasters remain up to date. Moreover, it remains in line with global scientific knowledge, thereby strengthening teaching on the subject in the universities and faculties of the National Health

System. Instructors have also delivered postgraduate training in several Caribbean and Central American countries, helping CLAMED to achieve affiliation with the Medical University of Havana, which has the overall aim of developing postgraduate education in all areas of the National Health System (Sánchez et al. 2016).

The dissemination of information outputs, scientific materials, statistics and news items of interest, virtual and on-site scientific events, as well as the building of websites and other virtual spaces has been facilitated by a number of factors. These have included, among others, the development of teaching with a focus on health and disasters; the high level of education achieved by the population in general; the universalization of higher education; the creation of the data transmission network INFOMED as part of a project that has led to greater collaboration, open access for publications and information sources, as well as web-based learning and online research (Alfonso Sánchez 2017); and the development of medical libraries across the National Health System. Such initiatives have resulted in the publication of the book *Gerencia de Desastres en Cuba* (Disaster Management in Cuba*)*, containing more than 500 articles. Moreover, more than 450 unedited works by Cuban authors have been collated in a ten-volume collection entitled *Salud y Desastres: Experiencias Cubanas* (Health and Disasters: Cuban Experiences) in digital and hard copy, available throughout the country (Borroto O'Farrill and Rodríguez Luis 2009; Céspedes Mora 2007; Colectivo de Autores 2015a; Colectivo de Autores 2015b). All of the above allows for improved communication, information management, and knowledge.

CLAMED's methodological endeavors have always been indelibly linked to its teaching activities, achieving significant results through its multidisciplinary and intersectoral work. A positive outcome of the coordination and integration of such efforts is knowledge of the hazards, vulnerabilities, and risks within the sector and the support it offers to MINSAP in drawing up legal documents which comply with the laws, decree laws, decrees, and resolutions of the Cuban state.

A national strategy has also been devised for the implementation of guidelines concerning mental health in disaster situations. In place since 2008, this strategy aims to minimize the risks and vulnerabilities linked to human subjectivity and the different ways in which it perceives the objective reality of a given situation, with the wider objective of increasing protective and resilience factors. A psychological preparedness program has subsequently been incorporated, aimed at the main specialist bodies within the sectors that respond to sudden events. In addition, a disaster-reduction strategy for chemical and radiological accidents has been developed, with a view to improving the intervention of health services at all levels, to detecting existing gaps, and to increasing the efficacy of preparation and response. Equally, a practical guide for food and nutrition management of vulnerable groups in emergency situations has been compiled.

Furthermore, since 2005, impetus has been given to the Hyogo Framework for Action. This impetus is seen in the application of vulnerability assessment methodologies proposed by PAHO/WHO and supported by Cuban experts, including the validation of the Hospital Safety Index, a method of assessing hospital security. The latter assessment led to the training of investors, drafting technicians, architects, engineers, medical personnel, and managers linked to the investment process, as well as the maintenance and distribution of health services (Jiménez Acosta et al. 2014; Lorenzo et al. 2012; Mesa Ridel et al. 2011; Navarro et al. 2014; Reyes and Mesa 2011). Additionally, methodologies have been developed for social communication between professionals and the community, and also for improving the culture of disaster prevention by following the strategic guidelines put forward by the ISDR (Roca et al. 2012; Rodriguez et al. 2009; Villalobos Mora 2000).

CLAMED's continually developing focus on health and disasters has proven to be an effective step forward for health sector training processes. It has improved the performance level of professionals, experts, teaching staff, and management, in that the competencies acquired have facilitated successful outcomes before, during, and after disaster events in Cuba, as well as in relation to international medical collaboration. Education in the workplace and the fusion of theory and practice constitute the pillars of these teaching processes, brought to life by the combination of study and work in health services, the interaction between the medical universities and the community, and the network of teaching and research institutions (Vela Valdés et al. 2018). Through applying scientific methods to solve practical health and educational problems, the blend of teaching and medical care has also boosted scientific research as an integral part of postgraduate training and extracurricular activity.

THE CHALLENGES OF CONFRONTING CLIMATE CHANGE

Climate change is the most pressing global hazard in today's world. While it may have some beneficial effects on human health, the international scientific community now believes that the repercussions of climate change regarding health will be largely adverse (IPCC 2014; McMichael 2003). At the regional level, there is evidence of changes and projected impacts that exert a direct or indirect influence on individual and population health, as well as that of health institutions (Centella et al. 1997; Solomon et al. 2007; World Health Organization 2018).

The links between climate change and health form a convincing argument in favor of training professionals, in general, alongside plans for research that will evolve as evidence emerges and that will be consistent with local needs. This training must prioritize the control of climate-related diseases, health promotion and prevention at a local level, and the provision of medical care in a responsible and sustainable manner. Teaching about climate change offers a means of improving public and environmental health literacy, both of which are essential elements in building a climate-resilient society and health sector (Maxwell and Blashki 2016).

Climate change itself can be useful for enhancing learning and for improving the values, knowledge base, and quality of the professional individual. The latter will need to adopt new ways of living and specific methods of protection in order to educate others and apply sustainable, preventative measures grounded in a number of tenets: social, economic, and ecological principles; key concepts in human development; sustainability; quality of life; collective health; intercultural education; social cohesion; and population self-care (Sánchez Sanabria et al. 2013; Sauchay and Mesa 2015).

Cuba, as an island nation, cannot escape this reality. In 2017, the Cuban state approved a national plan titled *Tarea Vida* (Project Life), which includes five strategic actions for reducing greenhouse gas emissions and vulnerabilities in the short, medium, long, and very long-term (from the year 2020 to 2100). All of its actions are closely tied to the 2030 Agenda and the Sustainable Development Goals (SDGs) (República de Cuba 2017a, 2017b).

MINSAP's Plan of Action involves the areas of environmental health, epidemiology, vector control, science and technological innovation, medical education, communication, and investment, as well as all of the country's universities and research centers. These actions all form part of its efforts to contribute to *Tarea Vida* through sustainable health sector activities from a scientific, health, and environmental perspective. In terms of climate change, postgraduate education must focus on two main imperatives: reversing negative trends through mitigation and reducing vulnerability to extreme events via adaptation and prevention of environmental contingencies.

The potential for tackling climate change in Cuba's medical science universities has increased following the incorporation of themes of "adaptation" and "mitigation" into the plans for environmental education. Studies exist on the hazards, vulnerabilities, and risks posed by disasters, the results of which are made available to the country's provinces. These studies employ methodologies suited to all types of hazards, which enable a conceptual understanding of their key components and, in turn, the necessary resilience-building processes.

Additionally, the curricula of the various medical sciences degree programs address issues that guarantee knowledge acquisition and the development of competencies. The curriculum favors active student participation in tasks aimed at combatting climate change, the development of a sense of responsibility regarding the ecosystem, and the carrying out of activities concerning environmental education, promotion, and care within communities. This approach contributes to improving risk perception, increasing knowledge, and fomenting an energy and water-saving culture with respect to human consumption.

The training of health sector professionals in disaster mitigation must also foster climate resilience in health institutions and the reduction of carbon emissions through projects that complement programs and plans. As result, the indirect benefits (co-benefits) for human health will be maximized. These benefits include saving water, energy efficiency, the prevention and reduction of fires, the rational use of medicinal products, the development of renewable energy sources, the adequate management of sewage, chemical and biological waste, and reforestation in health facilities. Special emphasis must be placed, however, on acquiring the necessary tools that will enable postgraduate education to exert a greater influence over climate change adaptation in Cuba. This adaptation should be viewed as a process that encourages the design, implementation and evaluation of strategies, policies, and measures to reduce the real and potential impacts of climate change. It should also take advantage of its opportunities, including the experiences most conducive to the development of vulnerabilities, interventions to reduce these, and risk perception (Ebi and Semenza 2008; Tschakert and Dietrich 2010).

The Cuban health sector continues to adhere to the underpinning aims of improving population health and increasing quality and therefore satisfaction with its service provision, as well as making the system more efficient and sustainable and guaranteeing its development (Ministerio de Salud Pública 2010). The sector's efforts regarding climate change mitigation and adaptation, however, must find a more integral space in their recently applied strategy for professional training and development, with a view to implementing systemic management related to disaster risk reduction. Such management will improve structures and, in turn, strengthen essential public health functions. Consistent with the vision, mission, and practices of sustainable development, this can all be achieved through forward-looking resilience-building and the necessary transformative potential of young scientists.

CONCLUSIONS

The Cuban health sector's postgraduate education strategy has in large part determined the achievement of the objectives set for the management of disaster risk reduction and the tackling of climate change. It attaches great importance to updating curricula in line with the population's health problems, particularly with respect to the following: the framework of development for primary health care; the incorporation of teaching materials concerning the prospective adaptation to climate change; the updating of the medical teaching–research–care link through systemic management involving health institutions; and the bridging of gaps in understanding, addressing, and transforming vulnerabilities and capacities for adaptation. All of these elements allow education to contribute to Cuba's strategic, programmatic, interdisciplinary, intersectoral, and local approach to disaster management. At the same time, they ensure the maintenance of the country's health indicators and the quality of, and satisfaction with, its health services, without neglecting the overall comprehensive, humanist, and solidarity-based nature of Cuba's health sector.

BIBLIOGRAPHY

Alfonso Sánchez, Ileana R. 2017. "XXV Anniversary of Infomed, the Cuban Telematic Net of Health." *Revista Cubana de Información en Ciencias de la Salud* 28 (3): 1–2.

Asamblea Nacional del Poder Popular. 1966. *Ley del Sistema de Defensa Civil de Cuba.* Accessed April 19, 2020. http://www.cubadefensa.cu/?q=ley-sistema-defensa-civil&b=d2.

Asamblea Nacional del Poder Popular. 1974. *La Ley No. 75 de la Defensa Nacional.* Accessed April 19, 2020. http://www.cubadefensa.cu/?q=ley75.

Bello Gutiérrez, Bruno. 2011. "La Medicina de Desastres, su inclusión en el programa de formación del Médico General Integral en Cuba." *Panorama Cuba y Salud* 6 (2–3): 26–31.

Bello Rodríguez, Berta, Esther G. Báez Pérez, Gilberto Sánchez Cruz, José Fernández Morín, Jana Fernández Alfonso, and Lenia Delgado Pérez. 2011. "La concepción integral de la salud: principio de la educación en ciencias de la salud." *Revista Médica Electrónica* 33 (6): 807–816.

Borroto O'Farrill, T., and Y. Rodríguez Luis. 2009. "La colaboración e intercambio de experiencias del Centro de documentación e información con instituciones del país y de la Región." In *Colectivo de autores. Salud y Desastres. Experiencias Cubanas*, Vol. I, 64. Havana: Editorial Ciencias Médicas.

Castellanos, A. 2012. "Los desastres y fundamentos de la protección contra los mismos." In *Colectivo de autores. Salud y desastres. Experiencias cubanas*, Vol. VI, 73–99. Havana: Editorial Ciencias Médicas.

Centella, A., L. Naranjo, L. Paz, P. Cárdenas, B. Lapinel, M. Ballester, R. Pérez, A. Alfonso, Cecilia González, M. Limia, and Marina Sosa. 1997. *Variaciones y cambios del clima en Cuba.* Havana: Centro Nacional del Clima, Instituto de Meteorología.

Céspedes Mora, Vivian María. 2007. "Los desastres, la información y el Centro Latinoamericano de Medicina de Desastres." *ACIMED* 16 (2). Accessed April

16, 2020. http://scielo.sld.cu/scielo.php?script=sci_arttext&pid=S1024-94352007000800006&lng=es.

Chávez Quintana, Pablo R. 2010. "Reseña histórica sobre la medicina veterinaria de desastres en Cuba." *Revista Electrónica de Veterinaria* 11 (3B): 1–4.

Colectivo de Autores. 2015a. "Salud y Desastres." In *Experiencias cubanas,* Vol. I-X. Havana: Editorial Ciencias Médicas.

Colectivo de Autores. 2015b. "Salud y Desastres." In *Experiencias cubanas*, Vol. Especial, 132. Havana: Editorial Ciencias Médicas.

Consejo de Estado. 1997. *Decreto-Ley No. 170 del Sistema de Medidas de Defensa Civil y normas constructivas*. Havana: Estado Mayor de la Defensa Civil.

Ebi, Kristie L., and Jan C. Semenza. 2008. "Community-Based Adaptation to the Health Impacts of Climate Change." *American Journal of Preventive Medicine* 35 (5): 129–139.

Évora Capote, Israel. 2012. "Reducción de riesgos de desastres y enfrentamiento al cambio climático." *Revista Congreso Universidad* 1 (1).

Fernández Valdés, María de las Mercedes, and Roberto Zayas Mujica. 2016. "Las competencias informacionales como determinante para el uso equitativo de la información científica y la tecnología en salud." *Bibliotecas Anales de Investigación* 12 (1): 90–97.

Ferreiro, Y., G. Mesa, Y. Sánchez, L. Sauchay, L. L. Gómez, and E. M. Roca. 2014. "Un quinquenio de experiencias en la formación de Postgrado en el Centro Latinoamericano de Medicina de Desastres. Año 2008–2012." In *Colectivo de autores. Salud y Desastres. Experiencias Cubanas*, Vol. VIII, 61–68. Havana: Editorial Ciencias Médicas.

Ferreiro Rodríguez, Y. 2015. "Rol de enfermería en el proceso de planificación y organización para la reducción de desastres en Cuba." In *Colectivo de autores. Salud y Desastres. Experiencias Cubanas*, Vol. Especial, 129–132. Havana: Editorial Ciencias Médicas.

Ferreiro Rodríguez, Yudix, Guillermo Mesa Ridel, Ricardo Pereda González, Dalilis Druyet Castillo, and Eva Caridad Apodaca Pérez. 2017. "Preparativos para la respuesta asistencia humanitaria ante conflictos armados, desastres y crisis complejas." *EDUMED*. Accessed April 14, 2020. http://edumed2017.sld.cu/index.php/edumed/2017/paper/view/265/0.

Gala, Ángela, María Eugenia Toledo, Yanisnubia Arias, Manuel Díaz González, Ángel Manuel Álvarez Valdez, Gonzalo Estévez, Rolando Miyar Abreu, and Gustavo Kourí Flores. 2012. "Implementación del Reglamento Sanitario Internacional en Cuba: Evaluación de capacidades básicas del sector salud en provincias seleccionadas." *Revista panamericana de salud pública* 32 (3): 207–216.

García Gómez, Vincente. 2016. "Sistema sanitario y preparación ante desastres." *La Universidad* (12): 128.

Guerrero Africani, Beatriz Helena. 2016. "La educación para la gestión del riesgo de desastres." Universidad Militar Nueva Granada. Accessed April 15, 2020. http://hdl.handle.net/10654/14755.

Hernández Roque, María de la Luz, Bárbara Toledo Pimentel, and Teresa Cabañin Recalde. 2014. "Currículo y actuación del Estomatólogo General ante situaciones de desastres." *EDUMECENTRO* 6 (1): 188–201.

IPCC. 2014. "Cambio Climático 2014: Impactos, Adaptación y Vulnerabilidad." Accessed March 15, 2020. https://www.ipcc.ch/pdf/assessment-report/ar5/wg2/ar5_wgII_spm_es.pdf.

Jiménez Acosta, Santa, Blanca Terry Berro, Armando Rodríguez Salvá, José Carrera Vara, and Guillermo Mesa Ridel. 2014. *Guía Práctica para el manejo alimentario nutricional de grupos vulnerables en situaciones de emergencias: Qué debemos identificar y qué hacer*. Havana: Ministerio de Salud Pública and UNICEF.

López Nicles, Roberto, Rolando Martínez Tamayo, and Andy López Utria. 2018. "La Universidad de Guantánamo en el fortalecimiento de las medidas de defensa civil en el malecón baracoense." *Revista EduSol* 18 (64): 126–140.

Lorenzo, A., C. Martinez, R. E. Ventura, and G. Mesa. 2012. "Lineamientos para la salud mental en desastres en Cuba." In *Colectivo de autores. Salud y Desastres. Experiencias Cubanas*, Vol. V, 3–10. Havana: Editorial Ciencias Médicas.

Marimón Torres, Nestor, and Evelyn Martínez Cruz. 2010. "Evolución de la colaboración médica cubana en 100 años del Ministerio de Salud Pública." *Revista cubana salud pública* 36 (3): 254–262.

Maxwell, Janie, and Blashki Grant. 2016. "Teaching About Climate Change in Medical Education: An Opportunity." *Journal of Public Health Research* 5 (1): 673.

McMichael, A.J. 2003. "Global Climate Change and Health: An Old Story Writ Large." In *Climate Change and Human Health: Risks and Responses*, edited by A.J. McMichael, D.H. Campbell-Lendrum, C.F. Corvalán, K.L. Ebi, A. Githeko, J.D. Scheraga and A. Woodward, 1–17. Geneva: World Health Organization.

Mejías Chao, Tania, Liliam María Mejías Chao, María Cristina Cabrera Rodríguez, and José Luis Amador Mejías. 2015. "La Medicina de Desastres en el plan de estudios del médico general." *Revista Educación Médica del Centro* 7 (3): 177–187.

Mesa Ridel, Guillermo. 2003. "Los desastres y la experiencia cubana en salud para su prevención." *Revista Cubana de medicina general integral* 19 (3).

Mesa Ridel, G., L. L. Gómez, L. Sauchay, G. Montalvo, P. Rodríguez, G. Reyes, and R. Monzote. 2011. "Instituciones de salud tras el paso de los huracanes Gustav e Ike. Resultados del trabajo de un equipo del CLAMED." In *Colectivo de autores. Salud y Desastres. Experiencias Cubanas*, Vol. IV, 219–227. Havana: Editorial Ciencias Médicas.

Mesa Ridel, Guillermo, Joel González García, María Cristina Reyes Fernández, Duniesky Cintra Cala, Yudix Ferreiro Rodríguez, and José Ernesto Betancourt Lavastida. 2018. "El sector de la salud frente a los desastres y el cambio climático en Cuba." *Revista panamerica de salud pública* 42 (24): 1–9.

Ministerio de Salud Pública. 2010. *Transformaciones necesarias en el Sistema de Salud Pública*. Havana: MINSAP.

Navarro, V. R., G. Mesa, A. Falcón, C. Llanes, and Y. Quintero. 2014. "Gestión para la reducción de riesgo de desastres en instituciones de salud; recomendaciones del taller de expertos realizado en Cienfuegos, 2012." In *Colectivo de autores. Salud y Desastres. Experiencias Cubanas*, Vol. VIII, 8–18. Havana: Editorial Ciencias Médicas.

Oramas González, René, Tamine Jordán Severo, and Norberto Valcárcel Izquierdo. 2013. "Competencias y desempeño profesional pedagógico hacia un modelo del profesor de la carrera de Medicina." *Educación Médica Superior* 27 (1): 123–134.

Paniagua Sergio. 1995. "Los desastres naturales y sus implicaciones en América Central." *Revista Geológica de América Central* 18: 107–112.

Quesada Romero, Reimundo. 2014. "Transformaciones de la educación superior en Cuba y el modelo pedagógico de la disciplina preparación para la defensa." *Congreso Universidad* 3 (2).

Ramos Guadalupe, Luis Enrique. 2009. *Huracanes: Desastres Naturales en Cuba*. Havana: Editorial Academia.

República de Cuba. 1999. *Decreto No. 262. Reglamento para la compatibilización del desarrollo económico-social del país con los intereses de la defensa*. Accessed August 30, 2017. http://www.mvd.sld.cu/base_legal/Decreto%20%20262%20Com patibilizaion%20de%20desarrollo.pdf.

República de Cuba. 2017a. "Bases del plan nacional de desarrollo económico y social hasta el 2030: Visión de la nación, ejes y sectores estratégicos."

República de Cuba. 2017b. "Tarea Vida." Accessed September 21, 2017. https://www.ecured.cu/Tarea_Vida.

Reyes, M.C., and G. Mesa. 2011. "Instituciones de salud seguras. Experiencias alcanzadas. Cuba 2009." In *Colectivo de autores. Salud y Desastres. Experiencias Cubanas*, Vol. IV, 29–34. Havana: Editorial Ciencias Médicas.

Roca, E., G. Mesa, and G. Montalvo. 2012. "Prácticas comunitarias en la gestión de reducción de riesgo." In *Colectivo de autores. Salud y Desastres. Experiencias Cubanas*, Vol. V, 73–77. Havana: Editorial Ciencias Médicas.

Rodriguez, Y., A. Lorenzo, G. Mesa, and C. Salabarría. 2009. "Desarrollo de capacidades locales en situaciones de desastres. Isla de la Juventud, una experiencia para compartir." In *Colectivo de autores. Salud y Desastres. Experiencias Cubanas*, Vol. III, 123–128. Havana: Editorial Ciencias Médicas.

Sánchez, Ileana. 2000. *CRID: Fuente de información sobre desastres en América Latina y el Caribe*. Caracas, Venezuela: Fundación de Edificaciones y Dotaciones Educativas.

Sánchez Sanabria, Marina, Mariemma Socarrás Vega, Francia Esther Herrera, Lina Tatiana Marín Picón, and Diomedes Artunes Noriega Galindo. 2013. "Mitigar consecuencias del calentamiento global y efecto invernadero: Reflexiones para la formación en salud." *Hacia la promoción de la salud* 18 (2): 110–122.

Sánchez Tarragó, Nancy, and Yudaimys Arriola Martínez. 2014. "La vigilancia epidemiológica internacional en Cuba 2009–2011." *Revista cubana de salud pública* 40: 317–330.

Sánchez, Y., A. Lorenzo, G. Mesa, M. Del Huerto, and L. Sauchay. 2016. "Propuesta metodológica para el posgrado internacional en Salud y Desastres. Centro Latinoamericano de Medicina de Desastres." In *Colectivo de autores. Salud y Desastres. Experiencias Cubanas*, Vol. IX, 7–11. Havana: Editorial Ciencias Médicas.

Sauchay, L., and G. Mesa. 2015. “Acercamiento al cambio climático y el sector salud, un reto para los países subdesarrollados.” In *Colectivo de autores. Salud y Desastres. Experiencias Cubanas,* Vol. Especial, 16–22. Havana: Editorial Ciencias Médicas.

Sauchay, L., G. Mesa, Y. Sánchez, P. Rodriguez, and M. Reyes. 2016. “Plan de reducción de desastres: una herramienta eficaz para la reducción de riesgos.” In *Colectivo de autores. Salud y Desastres. Experiencias Cubanas*, Vol. VIII, 1–7. Havana: Editorial Ciencias Médicas.

Solomon, Susan, Dahe Qin, and Martin Manning. 2007. “Resumen Técnico.” In *Cambios Climáticos: Base Física de la Ciencia*, edited by Susan Solomon, Dahe Qin, Martin Manning, Melinda Marquis, Kristen Averyt, Melind M.B. Tignor, Henry LeRoy Miller, Jr., and Zhenlin Chen, 77. Cambridge: Cambridge University Press.

Teja Pérez, Julio, and Guillermo Mesa Ridel. 2017. “Formación de Recursos Humanos en la gestión para la reducción de riesgos de desastres del Sistema de Salud y la Colaboración Internacional.” *Infodir* 13 (24): 78–85.

Tschakert Petra, and Kathleen Ann Dietrich. 2010. “Anticipatory Learning for Climate Change Adaptation and Resilience.” *Ecology and Society* 15 (2).

Vela Valdés, Juan. 2017. “Formación de médicos para los servicios de salud en Cuba 1959–2014.” *INFODIR*. Accessed April 20, 2020. http://revinfodir.sld.cu/index.php/infodir/article/view/405.

Vela Valdés, Juan, Ramón Syr Salas-Perea, María Luisa Quintana-Galende, Nayra Pujals-Victoria, Jorge González-Pérez, Lázaro Díaz-Hernández, Lizette Pérez-Perea, and María J. Vidal-Ledo. 2018. “Formación del capital humano para la salud en Cuba.” *Revista panamericana de salud pública* 42: 1–8.

Vidal Ledo, María, Bertha Fernández Oliva, Ileana R. Alfonso Sánchez, and Ileana Armenteros Vera. 2004. “Información, informática y estadísticas de salud: un perfil de la tecnología de la salud.” *ACIMED* 12 (4). Accessed April 15, 2020. http://scielo.sld.cu/scielo.php?script=sci_arttext&pid=S1024-94352004000400008&lng=es&tlng=pt.

Villalobos Mora, Margarita. 2000. “Estrategia de Comunicación para construir una cultura de prevención.” *Revista EIRD Informa.* Accessed April 18, 2020. http://cidbimena.desastres.hn/docum/crid/EIRDInforma/esp/revista/No2_2001/Pagina10.htm.

WADEM. n.d. “World Association for Disaster and Emergency Medicine.” Accessed April 18, 2020. https://wadem.org/about/association-overview/.

World Health Organization. 2018. *COP24 Special Report: Health and Climate Change*. Geneva: World Health Organization.

Chapter 8

People Power

Cuba's Path to Effective Disaster Management

Lauren Collins

A guiding principle of the Sendai Framework for Disaster Risk Reduction is that it is the primary responsibility of the state to put appropriate measures in place; it also emphasizes the need for coherence between local and national frameworks for safeguarding lives, livelihoods, and health (Sendai Framework 2015). Since 1959, the development of Cuba's national framework of laws, institutions, and public policies has been an incremental process in step with the capacity of the island's communities to participate in, and contribute to, disaster management measures. Many researchers seeking to understand Cuba's renowned success in this endeavor have acknowledged that it is underpinned by its socio-political model and high educational standards (Kirk 2017; López-Marrero and Wisner 2012; Nijhoff 2005; UNDRR 2004). Castellanos Abella and Wisner (2019, 30) note that "the evolution of the civil defense system is a mirror of increasing sophistication of local government and the increasing educational level of the population" and that the multilateral framework within which natural hazards governance is situated is the product of continuous improvement spanning sixty years. Initiated by the national leadership in the early years of the Revolution, and referred to in Cuba as *perfeccionamiento,* continuous development is achieved by processes of deliberation, experimentation, review, consultation, and discussion, which are embedded in Cuban political culture. Both the education system and the political-administrative apparatus have undergone (and continue to undergo) cycles of *perfeccionamiento.*

The purpose of this chapter is to explore more fully the relationship between the evolution of local government and education policy and that of the Civil Defense (Defensa Civil, or DC) system. It first offers an overview of the current political administrative system, before examining how the evolution of local structures, designed to involve communities in the business of local government, has promoted a culture of participation and solidarity, resulting in institutional and organizational infrastructures which are critical to effective disaster management. It then explores the way in which education policies have created a population capable of responding to, and participating in, disaster prevention measures, while also producing a scientific and technological base that can confront a wide range of natural hazards. The final section demonstrates, as Castellanos Abella and Wisner (2019) have asserted, that the evolution of the DC system has occurred hand-in-hand with an increasingly participatory local polity and rising educational standards. In turn, it explains why disaster and risk reduction management cannot be regarded as a system designed by experts in ivory towers and deployed in times of emergency by "officials with evacuation plans" (Nijhoff 2005, 13).

THE CURRENT POLITICAL ADMINISTRATIVE SYSTEM

From a liberal democratic standpoint, Cuba is seen as a highly centralized polity, with an overbearing national government, little or no place for localism, and few, if any, powers devolved to the municipalities. One consequence of this view is that, despite the large body of literature produced in response to the Revolution, local government and its participatory mechanisms have received scant attention. Yet, any examination of Cuba's path to effective disaster management must consider the evolution of Cuba's local polity and the political culture that it generates, reflects, and sustains.

Cuba's political administrative system has three tiers—national, provincial, and municipal—and comprises three components: Popular Power (Poder Popular), the mass membership organizations, and the Cuban Communist Party (Partido Comunista de Cuba, or PCC). Popular Power consists of a national assembly and 169 municipal assemblies. The latter are composed of elected delegates, who in turn elect a president and at least one vice president from among their number. Responsibility for day-to-day oversight of many services and production centers has been devolved to the municipal assemblies. Each municipality is sub-divided into smaller areas, each served by a Popular Council (Consejo Popular). These Councils are formed by delegates elected to represent their community in the municipal assembly and representatives of the local branches of the mass membership organizations, together with others that the local councils consider helpful for improving services in the communities under their jurisdiction. These individuals may include, for example, teachers, factory or farm managers, community health professionals, architects, and social workers. It is important to note that Popular Councils play a significant role in disaster management.

Between the national and municipal levels of government, there are fifteen provinces, each with a full-time governor who is supported by a Provincial Council (Consejo Provincial), which consists of the presidents and vice presidents of all the province's municipal assemblies. The provincial governor and council act as a link between the municipalities and central government and coordinate the work of the municipal assemblies, including that related to DC. The National Assembly, Cuba's parliament and only legislative body, has 600 seats, up to half of which must be elected municipal delegates, creating a link between national and local government.

The four largest mass membership organizations are the Committees for the Defense of the Revolution (Comités de Defensa de la Revolución, or CDR), the Federation of Cuban Women (Federación de Mujeres Cubanas, or FMC), the Workers Central Union of Cuba (Central de Trabajadores de Cuba, or CTC) and the National Association of Small Farmers (Asociación Nacional de Agricultores Pequeños, or ANAP). Established in 1960, the CDR and the FMC are both products of the Revolution, while the CTC, an umbrella organization representing Cuban trade unions, was founded in 1939. ANAP represents the interests of private farmers and also pre-dates the Revolution. Most Cubans hold membership of one, or more, of these organizations, all of which have local branches and are an important mechanism for citizen participation.

All four of the mass organizations pre-date the PCC, Cuba's only political party, established in 1965. According to the 2019 Constitution, it is the "superior driving force of the society and the State," although it plays no role in the electoral process (Constitution of the Republic of Cuba 2019). The development of the national framework for disaster management has been driven by the PCC and the national assembly.

BUILDING ORGANIZATIONAL AND INSTITUTIONAL INFRASTRUCTURE

In January 1959, when Fidel Castro's Rebel Army succeeded in overthrowing dictator Fulgencio Batista, the Cuban state apparatus collapsed almost completely, leading to "something of a *tabula rasa*" (Kapcia 2008, 24). Local government had to be built from scratch. During the Revolution's first two years, provinces and municipalities were administered by government-appointed commissioners. The long process of building collective community action began in late 1961, when Coordination, Implementation and Inspection Boards (Juntas de Coordinación, Ejecución e Inspección, or JUCEI) were established across the country. They were to be the vehicle through which citizens could participate in local affairs via the inclusion of the mass membership organizations within the assemblies, boards, and executive committees of JUCEI, introduced in every province, region, and municipality (Collins 2017).

After a period of experimentation and pilot schemes, 1965 saw the establishment of regional and municipal local administrations (Administraciones Locales) to administer public service delivery and manage local commercial and production activities. The following year, direct elections took place in popular assemblies in neighborhoods and workplaces. 10,887 delegates were elected to the executive committees of Local Power (Poder Local) (Collins 2017), which were created in each of the then 407 municipalities. Composed of ten elected delegates and the municipal general secretaries of each of the mass membership organizations, the role of these Local Power committees was to hold the officials of the local administrations to account and attend to problems raised by their electors. With the support of the local branches of CDR, Local Power continued to function until 1976 (Booth 1973).

The active involvement of branches of the mass membership organizations in these early forms of local government contributed to building the capacity for collective and cooperative action at a community level. It also initiated the forging of a political culture of what may be termed "civic responsibility," although the term "social solidarity" perhaps more closely reflects the wider sense of collective responsibility.

Popular Power

Drawing on the experience of JUCEI and Local Power, and following the approval by referendum of the first Constitution of the Revolution in February of 1976, new political and administrative structures, alongside a system of elected government bodies at the national, provincial, and municipal levels, were implemented. The number of provinces was increased from six to fourteen and the 58 regions and 407 municipalities were reconfigured to form 169 municipalities, eliminating the middle tier of regions completely. The mass movement organizations and the PCC were restructured to mirror the new political administrative arrangements. The organizational alignment of the three components of Cuban polity—Popular Power, the mass membership organizations, and the PCC—radically improved coordination and communication, both vertically and horizontally, between national and local socio-political institutions. This alignment laid the basis of the institutional infrastructure which underpins the success of the current disaster management system.

The first delegates to the municipal assemblies of Popular Power were elected in October 1976. Approximately, one in every 1,200 people served as delegates at any one time, and, with an average of 50 percent new delegates returned at each election, a significant number of citizens have experienced incumbency (Roman 2003). Delegates must be a resident in the area they represent, receive no remuneration for their role, and remain in their usual employment. Brief details of the nomination and election process of municipal delegates are given here to illustrate the relationship between communities and their representatives, which impacts significantly on the effective implementation of disaster management measures.

Each municipality is divided into "wards," known as *circunscripciones*, of around 1,000 residents, each of which elects a delegate to the municipal assembly. Candidates are nominated in small neighborhood meetings convened across each *circunscripción*. At each meeting, neighbors nominate people whom they consider would make good delegates and following a general discussion, one nominee is chosen. Nominees from each of the meetings appear as candidates on the ballot paper on election day, when electors vote for one candidate to represent their *circunscripción* at the municipal assembly. The local method of nomination and election means that most candidates are known to the electors, and, more importantly, they have been chosen by, and from, the community. Delegates must be accessible to their electors through weekly "office hours" and must hold public accountability sessions at least every six months, where they report on the work of the municipal assembly, and take questions, suggestions, and complaints, known as *planteamientos*. All *planteamientos* are meticulously noted and must be responded to. The role of the municipal delegate is an important one in Cuba's dispersed disaster management system.

In December 1975, the first PCC Congress recommended that Popular Power be subject to continuous review (i.e., *perfeccionamiento*). Over the next decade (1976–1986), auxiliary commissions at the national, provincial, and municipal levels facilitated deliberation, experimentation, review, consultation, and discussion (Collins 2017). Through this process, it became apparent that delegates in more isolated areas often experienced difficulty in traveling regularly to administrative offices to respond to *planteamientos*.

As part of *perfeccionamiento*, the first official experiments with Popular Councils began in the late 1980s. In 1986, Law 56 provided the legal basis for their creation; subsequently, in 1988, 225 Popular Councils were formed (García Brigos 1996). They were incorporated into the Popular Power structures through constitutional reforms in 1992; in early 1993, Popular Councils sprung up in every municipality. Each council is made up of delegates in contiguous *circunscripciones*, together with representatives from mass membership organizations and other local actors. Popular Councils support the work of the delegates and connect communities more closely to local government. In times of emergency, for example hurricanes or flooding, every Popular Council becomes a Defense Council (Consejo de Defensa), adding a further dimension to what Castellanos Abella and Wisner (2019) describe as a "multilateral framework."

The story of Cuban polity since 1959 is one of a gradual process of decentralization. The structures and practices, which have evolved under the revolutionary government, are interwoven and overlapping, providing the population with multiple opportunities for participation in their communities and shaping the relationship between citizen and government.

BUILDING AN EDUCATIONAL AND SCIENTIFIC BASE

The determination to escape under-development has shaped Cuba's education policies. In 1959, the new leadership faced high levels of illiteracy, a lack of administrative and managerial experience, and a scarcity of scientific, technological, and engineering skills. The 1953 census is illustrative of this, concluding that in many rural areas, illiteracy was as high as 41.7 percent (Seers et al. 1964). By 1958, only 49.3 percent of all children attended primary school, while only 12 percent received any form of secondary education (Ibid.).

In particular, there was scant teaching of science, engineering, and technology. These problems were exacerbated by the exodus of 110,000 skilled, professional, and semi-professional Cuban and U.S. citizens during 1959 and 1960 (Olsen and Olsen 1995). They included technicians, managers, experienced administrators, and almost half of school and university teachers (Fitzgerald 1994).

Primary Education for All

During the Revolution's first decade, a key priority was increasing access to education (Lutjens 2000), which required eradicating illiteracy, establishing schools, training teachers, and raising the educational standards of working adults. In 1959 alone, 15,000 schools were opened. The Literacy Campaign of 1961, the "Year of Education," reduced the illiteracy level from 23.6 percent to 3.9 percent (Canfux and Mateja 1981). Following the campaign, the Department for Adult Education was created, establishing an education system for working adults parallel to that which served children and young people (Fitzgerald 1994). A follow-on program (*Seguimiento*) for newly literate adults was launched in February 1962.

To educate working adults to a level equivalent to completing primary school, the Battle for the Sixth Grade was launched in 1964. The program ran for two decades and was supported by the trade unions, ANAP, the FMC, and the CDR, all of which promoted the new educational opportunities, identified people who could teach others, and provided rooms and spaces where teaching could take place. Half-a-million adults who had learned to read and write during the Literacy Campaign obtained Sixth Grade certification by 1967 (Canfux and Mateja 1981). The number of teachers in schools rose considerably in the first decade, as did the number of children attending primary school. It was at this point that materials on disaster preparedness were first introduced.

Perfeccionamiento

To keep pace with the need for qualified schoolteachers, especially at the secondary school level, 95 technical and professional education centers were opened in 1971 (Fitzgerald 1994). By 1972, many adults who had begun their educational journey with the Literacy Campaign or the Battle for the Sixth Grade were now ready to attend secondary school, college, or university (MacDonald 2009). The demand for teachers at all levels, and the insufficiency of scientific and technical skills, were becoming urgent matters. The entire education system had been developed rapidly during the first decade, under difficult circumstances, and was thus in need of a *perfeccionamiento* process.

In early 1971, thousands of meetings were held throughout Cuba, facilitated by the National Union of Educational, Cultural, and Scientific Workers and attended by almost every educator in the country. The discussions and debates were followed by the First National Congress of Education and Culture in Havana, which mandated the Ministry of Education (Ministerio de Educación or MINED) to carry out a diagnostic study of the entire education system. The outcome was a resolution on education, approved at the First Congress of the PCC in 1975. Work then began in line with the comprehensive Plan to Perfect and Develop the National Education System (*Plan de Perfeccionamiento y Desarrollo del Sistema Nacional de Educación de Cuba*) (Fitzgerald 1994).

Decentralization of the administration of schools was made possible by the new provincial and municipal assemblies elected in 1976. The national curriculum, under the purview of MINED, was reorganized and rationalized. Teacher training was improved and expanded, particularly at the secondary level, to match the rapid growth in student enrolment (Lutjens 2000). The Battle for the Sixth Grade was declared complete in 1980 and in 1985 the Battle for the Ninth Grade began. Improvements in the quality of teaching in secondary schools fed through to higher education; enrolments increased by 583 percent between 1970 and 1984 (Fitzgerald 1994). The number of higher education institutions also grew from three in 1959 to 46 in 1984.

Creating a Scientific Base

The changes in higher education after 1975 are rooted in the 1962 university reform, which attempted to bolster the weak post-graduate and research sector by introducing student scholarships on a large scale. This led to a rise in the number of trained professionals, including researchers, within an education system increasingly orientated toward the sciences, engineering, medicine, and economics (Chávez Rodríguez 2011). In 1964, Ciudad Universitaria José Antonio Echeverría (CUJAE) was founded, specializing in engineering and technology, followed by the National Center for Scientific Investigations in 1965. Under the auspices of the Ministry of Public Health (Ministerio de Salud Pública, or MINSAP), in 1963 thirteen national institutes for studying the sciences were created (Majoli Viana 1999). The urgent need to bring healthcare to the whole population acted as a spur for university reform, which, by increasing the education and training of medical staff, laid the basis for Cuba's universal health system. The *perfeccionamiento* process saw the opening of country-wide medical training centers, enabling students to train, and subsequently practice, in their home provinces (Morales et al. 2008).

Higher education was key to making good the intention to increase the study of science, engineering, technology, and agronomy necessary for overcoming under-development. Thanks to the emphasis on teaching science to children and to adults in continuing education, the take-up of subjects within the sciences at universities increased, allowing Cuba to develop a scientific base. The Lenin School, opened in 1974, was the first Pre-university Vocational Institute for the Study of the Exact Sciences, and similar institutions were established in all provinces after 1976 (Fitzgerald 1994).

Throughout the 1990s, higher education enrolments, including postgraduate studies and research, continued to increase. The Ministry of Science, Technology and Environment (Ministerio de Ciencia, Tecnología y Medio Ambiente, or CITMA) was created in 1994 to promote policy and research in science and technology. By the end of the 1990s, there were over 600,000 graduates and more than 221 research centers catering to 60,000 researchers (Lutjens 2000; Majoli Viana 1999). The expansion of the higher education sector enabled research to flourish in the fields of meteorology, geology, agricultural sciences, epidemiology, computational science, and many others applicable to disaster risk reduction.

In 2009, Municipal University Centers (CUM) were established in every municipality. These offered postgraduate courses and research opportunities to working graduates, as well as advanced training to professionals, public officials, and teachers, in addition to contributing to local development (Hernández Gutiérrez et al. 2013). Municipal governments are required by law to adopt sustainable local development plans, which must include the mitigation of risks associated with natural hazards. It is usual, therefore, for CUM research programs to be coordinated with such plans, and hence to contribute to local disaster management.

The role that education has played in Cuba's path to effective disaster management is manifold. As acknowledged in the Education for All Global Monitoring Report of 2005 (UNESCO 2004), universal free education has resulted in high levels of education in the general population. This increase is beneficial in terms of people's response to government messaging about hazards and rational thinking regarding related risks. Consequently, it has also helped to undermine suspicion and antipathy toward science and technology, attitudes which were prevalent prior to the Revolution (Leiner 1981). School children were taught about the causes of extreme weather events, and what to do when they occur, while collective effort and mutual support were promoted in educational settings (Breidlid 2007; Sandoval Guzmán 2014). The free healthcare system, which takes a preventative approach in relation to natural hazards, would not have been possible without sustained investment in education. Moreover, the university system has enabled Cuba to establish a network of scientific research institutes and trained specialists in a wide range of subjects relevant to disaster risk reduction.

DISASTER RISK REDUCTION AND MANAGEMENT

In 1963, Hurricane Flora made landfall in Cuba killing more than 1,200, destroying 30,000 homes and causing an estimated $500 million of damage (Pérez 2009, 5). The disaster is generally regarded as the catalyst that set Cuba on the path to effective disaster management, although its origins can be traced back to 1962. That year, the Military Organization of Industries (Organización Militar Industrial, or OMI), established in 1959 to protect industry from U.S. military aggression and counter-revolutionary espionage, became the Central Headquarters of Popular Defense. Its creation marked the beginning of a nationwide approach to disaster management (Castellanos Abella and Wisner 2019). The existence of JUCEI in the provinces, regions, and municipalities formed an organizational base to respond to emergencies and allowed for the launch of a national Popular Defense network.

Yet, Hurricane Flora demonstrated that the Popular Defense was not necessarily able to cope with all challenges, particular meteorological ones. In the immediate aftermath, there began a nationwide program of dam-building and the modernization of the meteorological service. More significantly, however, for the future of effective disaster management, the catastrophe prompted a period of *perfeccionamiento*, culminating in Law 1194 in 1966, the year in which Local Power was established. The new law led to the creation of the National Civil Defense (Estado Mayor Nacional de la Defensa Civil, or EMNDC, or DCN). DC Councils, responsible for planning for the protection of the population and the economy in the event of natural disasters (Cuba Defensa n.d.), were introduced in each province. Law 1194 also mandated the preparation of evacuation plans, the provision of transportation, food and water supplies, planning for continuity of production, and the organization of warning systems, the latter made more effective by the creation of the Institute of Meteorology (Instituto de Meteorología, or INSMET) in 1965.

Law 1194 required all organizations to engage with DC tasks and appoint coordinators with specific responsibilities (Castellanos Abella and Wisner 2019). These bodies would certainly have included the new local administrations and the mass membership organizations. Local power delegates provided a link between local communities and the new DC Councils, creating an avenue for communication concerning the law's requirements and a means of involving the local population in disaster preparation.

The creation of Popular Power in 1976 coincided with Law 1316, which aligned the structures of the DC system with the new political administrative structures, the mass membership organizations, and the PCC. This alignment produced a coherent framework for DC activities at the provincial and municipal levels. DC Councils were established in all provinces and municipalities, and the presidents of the elected provincial and municipal assemblies became the presidents of the DC Council under their jurisdiction. This move was a significant step on the path to effective disaster management. The local delegate and the mass membership organizations, particularly the CDR and FMC, could disseminate information relevant to the residents in their neighborhoods, such as regarding decision-making and related actions, resulting in a greater degree of trust in DC measures, and therefore a high level of compliance in times of emergency.

Since 1986, a disaster and preparedness response exercise, known as *Meteoro*, has taken place annually over one weekend in every municipality, with the purpose of testing local disaster preparedness plans. Thanks to improved educational standards, municipalities could draw on local expertise to develop and evaluate such plans, which where, and remain, designed with input from mass membership organizations, elected delegates, community doctors, and other local actors, including directors of municipal health services and municipal DC Councils (Mesa 2008). Although hurricanes are the most common natural disasters in Cuba, disaster planning at the local level, and testing it through *Meteoro*, gave rise to a wider view of what constitutes a hazard. Municipalities differ in geographical characteristics, industrial and agricultural activities and weather patterns, and thus need to plan for different scenarios: industrial waste hazards, chemical spillages, landslides, wildfires, dam failure, earthquakes, droughts, and epidemics.

The advent of Popular Councils (Consejos Populares) in early 1993 was a further step in the decentralization of local government, allowing DC measures to be more finely tuned to the conditions and circumstances of local communities. Moreover, representatives from important service and production entities in the locality, and people with specialisms pertinent to the needs of the community, became formally active and involved in local government. In 2007, Popular Councils became "defense zones" in the event of an emergency or disaster, thus creating a unit of DC closer to the community. The capacity of local communities for collective action, built up over more than three decades, could now be leveraged for the purposes of honing Cuba's effective disaster management system, forming part of its "intangible qualities" (Thompson and Gaviria 2004, 7). Values of social solidarity generated through a participatory local polity became components of what Castellanos Abella and Wisner (2019) refer to as "the increasing sophistication of local government."

Drawing on the enhanced framework, which now encompassed the DC system, Law 170 was introduced in 1997. It established four stages of response in the event of a hurricane, including information, alert, alarm, and recovery (Kirk 2017). The staged approach gives time for plans to be put into action. A further decisive step toward the integration of disaster management and political and economic life was made with Decree Law 262 of 1999. This law mandated that all plans for investment and research be cross-checked by the DCN to avoid introducing additional risk factors as a result of development activities.

The extensive damage to infrastructure caused by Hurricanes Ivan and Charley in 2004 prompted another *perfeccionamiento* of the DC system, with the participation of 1,000 people from across the country. It identified shortcomings in local advance planning, and preparation and attention were turned to improving the development of local disaster risk reduction plans. A government directive ordered that every political administrative authority, and every commercial, production and service organization, create a disaster risk reduction plan. CITMA became responsible for training local experts in carrying out this task. The emphasis on reducing risk anticipated the Hyogo Framework 2005–2015, adopted by the Second World Conference on Disaster Reduction in 2005. This was also the year of the creation of defense zones at the level of the Popular Councils. These ameliorated the coordination of local human resources during the information and alert phases, while improved forecasting allowed the timing of stages to differ from place to place depending on the path of a hurricane or other major meteorological event. The meteorological service had benefited from a scientific and technological base supported by government education policies, which, in turn, had led to advancements in predicting the characteristics and likely paths of hurricanes since the days of Hurricane Flora. By 2010, there were sixty-eight weather stations throughout the island and a meteorological office in each of the provinces (Llanes Guerra 2010).

Surveillance of meteorological events and early warning systems were already in place when, in 2007, the first Risk Reduction Management Centers (RRMCs) were established in municipalities at high risk of hurricane damage. Directly responsible to municipal assemblies, the RRMCs provide information to aid decision-making in municipal assemblies, whose presidents lead the Municipal Defense Councils. The RRMCs produce, collect, and analyze a wide range of data to enable the production of risk reduction plans pertinent to the municipality's specific conditions and activities. This task requires the participation of local specialists from all sectors in the territory, for example: public health, agriculture, environment, sanitation, transportation, education, water resources, physical planning, housing, energy, universities, and research centers. Popular Councils and community leaders, such as those from the CDR and FMC, are also involved in the work of the RRMCs, while risk mapping is carried out by community members that "function as the mortar in Cuba's wall of risk reduction" (Thompson and Gaviria 2004, 31). Centers also create multidisciplinary teams to produce hazard, vulnerability, and risk studies (Peligro, Vulnerabilidad y Riesgo or PVR) for every Popular Council in their municipality (UNDP 2013). For each type of hazard, CITMA organizes national expert groups to combine specialist knowledge in drawing up guidelines and methodologies to support the production of PVRs (Castellanos Abella and Wisner 2019).

By 2019, there were 89 RRMCs in Cuba (Cuba Informe Nacional 2019). The network of RRMCs is an important component of the DC system, allowing information and warnings to be exchanged between the national and local levels. The center's ability to reduce vulnerability and increase community resilience depends on specialist expertise, and also the high standard of education of managers and administrators in all sectors of the economy and public services, as well as that of the general population.

High educational standards within any population depend on universal access. In Cuba, educating children about disasters, and how to prepare for them, is part of the wider formation of civic behavior and values, which stress collective responsibility and solidarity. From an early age, Cubans are taught about natural hazards, and children take home research tasks to learn about the risks they pose (Sandoval Guzmán 2014). Since 1967, disaster preparedness has formed part of the national curriculum and has remained under constant review. More recently, teaching material includes the study of vulnerability, resilience, climate change mitigation, sustainable development, risk management, early warning systems, and evacuation plans (GADRRRES 2017). Disaster studies is taught in the third and fifth grades of primary education, eighth grade of secondary education, and eleventh grade of high school. Schools also prepare risk maps and evacuation plans for disaster prevention (Selby and Kagawa 2012).

The MINED considers that educating children about hazards and disasters solely in a classroom setting does not alone foster broader community resilience. Since 1976, disaster risk reduction has thus included extracurricular activities in the community, in conjunction with the local branches of the mass membership organizations, with the aim of fostering participation in problem-solving and prevention (GADRRRES 2017). In many communities, children in the fifth grade are grouped into "guards" in their neighborhood, who are trained by the CDR to learn how to behave in the event of a hazard emergency (Pichler and Striessnig 2013).

CONCLUSION

The aim of this chapter has been to demonstrate how the steady advances in education—in terms of accessibility, increasing standards, and an emphasis on the sciences, combined with the evolution of a participatory local polity—have shaped Cuba's path to effective disaster management.

Coherence of national and local frameworks for effective disaster management, as advocated by the Sendai Framework, is evident in the Cuban approach to disaster and risk reduction and management. Although in the Cuban case the frameworks are tightly aligned, this has not prevented the development of plans and actions that reflect diverse conditions pertaining to specific communities. We have seen that Castellanos Abella and Wisner's (2019) "multilateral framework" has been made possible by the *perfeccionamiento* of local government and the education system. As the vice president of the Cuban Society of Hygiene and Epidemiology, Pedro Mas Bermejo, indicated, "Planning for natural disasters is integral to the political and economic life of Cuba, nationally and locally" (Mas Bermejo 2006). The need to "overlay" the established administrative structures with a specific structure at times of disaster is not necessary in Cuba nor is there a need for preparation activities to be conducted by external national or foreign organizations. The existing approach ensures continuity, speed, trust, and familiarity, all of which contribute to effective disaster management.

Cuba's effectiveness in disaster management is frequently attributed to a supposed authoritarian government that can "force" its population to comply with evacuation and pre-disaster activities (Aguirre 2005). This chapter has shown that Cuba's political culture, which promotes cooperation, social solidarity, and civic responsibility—all crucial elements in planning and responding to threats from natural hazards—stems from the participatory nature of the Cuban polity as much as the ideological context in which it takes place.

BIBLIOGRAPHY

Aguirre, B. E. 2005. "Cuba's Disaster Management Model: Should It Be Emulated?" *International Journal of Mass Emergencies and Disasters* 23 (3): 55–71.

Booth, David. 1973. "Neighbourhood Committees and Popular Courts in the Social Transformation of Cuba." PhD diss., University of Surrey.

Breidlid, Anders. 2007. "Education in Cuba—An Alternative Educational Discourse: Lessons to Be Learned?" *Compare* 37 (5): 617–634.

Canfux, Jaime, and John A. Mateja. 1981. "A Brief Description of the 'Battle for the Sixth Grade.'" *Journal of Reading* 25 (3): 226–233.

Castellanos Abella, Enrique, and Benjamin Wisner. 2019. "Natural Hazards Governance in Cuba." *Oxford Research Encyclopedia of Natural Hazard Science.* Accessed November 17, 2020. 10.1093/acrefore/9780199389407.013.238.

Chávez Rodríguez, Justo Alberto. 2011. "Education in Cuba between 1959 and 2010." *Estudos Avançados* 25 (72): 5–53.

Collins, Lauren. 2017. "The Dense Web: Local Governance and Popular Participation in Revolutionary Cuba." PhD diss., University of Nottingham.

Constitution of the Republic of Cuba. 2019. "Constitute: Cuba's Constitution of 2019." Accessed May 14, 2020. https://www.constituteproject.org/constitution/Cuba_2019.pdf?lang=en.

Cuba Defensa. n.d. "Se establece por ley el Sistema de Defensa Civil de Cuba." Accessed May 3, 2020. http://www.cubadefensa.cu/?q=ley-sistema-defensa-civil&b=d2.

Cuba Informe Nacional. "Sobre la Implementación de la Agenda 2030." 2019. Accessed May 29, 2020. https://foroalc2030.cepal.org/2019/sites/foro2019/files/informe _nacional_voluntario_de_cuba_sobre_implementacion_de_la_agenda_2030.pdf.

Fitzgerald, Frank T. 1994. *The Cuban Revolution in Crisis: From Managing Socialism to Managing Survival.* New York: Monthly Review Press.

GADRRRES. 2017. "Pillar 3: Students Leading Communities in Disaster Risk Reduction through Informal Education in Cuba." Global Alliance for Disaster Risk Reduction and Resilience in the Education Sector. Accessed May 5, 2020. https://www.preventionweb.net/files/61511_csspolicycasestudycubastudentsleadi.pdf.

García Brigos, Jesús P. 1996. "Los Consejos Populares: Origen, Evolución y Perspectivas." Accessed April 25, 2020. https://www.nodo50.org/cubasigloXXI/politica/gbrigos_consej_pop.pdf on 25.04.2020.

Hernández Gutiérrez, Dimas N., Berta Pich Herrera, and Francisco Benítez Cárdenas. 2013. "Los Centros Universitarios Municipales: Un Estudio de Caso de Responsabilidad Social Universitaria en Cuba." *Revista Congreso Universidad* 2 (2): 1–15.

Kapcia, Antoni. 2008. *Cuba in Revolution: A History Since the Fifties*. London: Reaktion Books.

Kirk, Emily. 2017. "Alternatives–Dealing with the Perfect Storm: Cuban Disaster Management." *Studies in Political Economy* 98 (1): 93–103.

Leiner, Marvin. 1981. "Two Decades of Educational Change in Cuba." *Journal of Reading* 25 (3): 202–214.

Llanes Guerra, José. 2010. *Cuba Risk Reduction Management Centres: Best Practices in Risk Reduction*. United Nations Development Programme.

López-Marrero, Tania, and Ben Wisner. 2012. "Not in the Same Boat: Disasters and Differential Vulnerability on the Insular Caribbean." *Caribbean Studies* 40 (2): 129–168.

Lutjens, Sheryl. 2000. "Educational Policy in Socialist Cuba: The Lessons of Forty Years of Reform." Construyendo Futuro Madrid: Foundation for Marxist Studies. Accessed May 14, 2020. http://www.angelfire.com/pr/red/cuba/educational_policy_in_cuba.htm.

MacDonald, Théodore H. 2009. *The Education Revolution: Cuba's Alternative to Neoliberalism*. Croydon, UK: Manifesto Press.

Majoli Viana, Marina. 1999. "Ciencia y Desarrollo en Cuba." Accessed May 30, 2020. https://www.paho.org/cub/index.php?option=com_docman&view=download&category_slug=antecedentes&alias=807-majoli-m&Itemid=226.

Mas Bermejo, Pedro. 2006. "Preparation and Response in Case of Natural Disasters: Cuban Programs and Experience." *Journal of Public Health Policy* 27 (1): 13–21.

Mesa, Guillermo. 2008. "The Cuban Health Sector and Disaster Mitigation." *MEDICC Review* 10 (3): 5–8.

Morales, Ileana Del Rosario, José A. Fernández, and Francisco Durán. 2008. "Cuban Medical Education: Aiming for the Six-Star Doctor." *MEDICC Review* 10 (4): 5–9.

Nijhoff, Martinus. 2005. "World Disasters Report." Centre for Research on the Epidemiology of Disasters and International Federation of Red Cross and Red Crescent Societies. Dordrecht, The Netherlands.

Olsen, James Stuart, and Judith E. Olsen. 1995. *Cuban Americans: From Trauma to Triumph*. New York: Twyne Publishers.

Pérez, Louis A. 2009. *Cuba: Between Reform and Revolution*. New York: Oxford University Press.

Pichler, Adelheid, and Erich Striessnig. 2013. "Differential Vulnerability to Hurricanes in Cuba, Haiti, and the Dominican Republic: The Contribution of Education." *Ecology and Society* 18 (3): 31.

Roman, Peter. 2003. *People's Power: Cuba's Experience with Representative Government*. Lanham, MD: Rowman & Littlefield.

Sandoval Guzmán, Javier. 2014. "Popular Power and Environmental Governance: The Cuban Approach to Natural Hazards and Disaster Risk Reduction." MA diss., Norwegian University of Life Sciences.

Seers Dudley, Richard Jolly, Andrés Bianchi, and Max Nolff. 1964. *Cuba: The Economic and Social Revolution*. Chapel Hill, NC: The University of North Carolina Press.

Selby, David, and Fumiyo Kagawa. 2012. *Disaster Risk Reduction in School Curricula: Case Studies from Thirty Countries*. Paris: UNICEF/UNESCO.

Sendai Framework for Disaster Risk Reduction 2015–2030. 2015. Accessed February 20, 2020. https://www.preventionweb.net/files/43291_sendaiframeworkfordrren.pdf.

Thompson, Martha, and Izaskun Gaviria. 2004. *Cuba, Weathering the Storm: Lessons in Risk Reduction from Cuba*. Boston, MA: Oxfam America.

UNDP. 2013. *Basic Training for Professionals on Risk Reduction Management Centres*. Havana: UNDP.

UNDRR. 2004. "Cuba: A Model in Hurricane Risk Management." Accessed May 19, 2020. https://www.undrr.org/news/cuba-model-hurricane-risk-management.

UNESCO. 2004. *The EFA Global Monitoring Report 2005*. Paris: UNESCO.

Chapter 9

Cuba's *Tarea Vida*

Sustainable Development and Combating Climate Change

Helen Yaffe

On April 12, 2020, Cuba's Institute of Meteorology (Instituto de Meteorología, or INSMET) registered the record-breaking temperature of 39.3° Celsius in Granma province, surpassing the record set just the preceding summer. While Cuba emits only 0.08 percent of world greenhouse gases, as a small island nation, it is disproportionately affected by climate change. The frequency and severity of extreme weather events (heat, cold, drought, floods, and hurricanes) are increasing, to the detriment of ecosystems, food production, and public health. Without action to protect the coast, the expected one-meter rise in sea levels will wipe out 122 Cuban coastal towns, affect water supplies, and destroy agricultural lands, submerging 2.3 percent of Cuba by 2050. However, action is being taken. In spring 2017, the Council of Ministers approved *Tarea Vida: State Plan to Confront Climate Change*. Within Cuba, *Tarea Vida*, which translates as "Project Life" in English, is conceived as more than a set of climate change mitigation and adaptation policies. It is regarded as a new basis for Cuban development, as part of a cultural change and a new articulation between the national and local government (Alfredo Carballo 2021). Indeed, it is integrated with the broader process of economic transformations and a phase of economic reforms involving decentralization of responsibilities, powers and budgets, and a strong emphasis on creating an environment to stimulate innovation and experimentation. Environmental concerns are integral, not marginal, to the country's national development strategy.

With a 100-year projection, *Tarea Vida* is currently the world's only truly long-term state plan to protect the population and physical environment from the imminent and escalating global threats presented by climate change.

Other countries' plans mostly focus on reducing carbon emissions while those addressing "climate resilience" are designed for a shorter term than Cuba's and the state is not the key actor. This is not by chance; it can be explained by several factors that can broadly be divided into those constituting necessity and those facilitating capacity.

From the perspective of necessity: First, centuries of colonial and then imperialist exploitation and the imposition of the agro-export model turned Cuba "from rainforest into cane field" as the expansion of the sugar industry led to chronic deforestation and soil erosion (Funes Monzote 2009)."[1] Redressing this historical legacy became part of the project for revolutionary transformation post-1959, which sought to break the chains of underdevelopment. Post-1959 activities that contributed to pollution and erosion have been gradually recognized and incrementally redressed, particularly from the 1990s. Second, the natural endowments of the Cuban archipelago of some 1,600 islands; Cubans are conscious of the island's extraordinary biodiversity and the global importance of its coastal resources. Third, Cuba's vulnerability to climate change as a small island nation and the threat to life on the island are described below.

In terms of capacity: First, Cuba's socialist system prioritizes human welfare, "and the social character of property facilitates environmental protection and the rational use of natural resources" (Houck 2000, 17). Thus, as law scholar Oliver A. Houck observed in his work on Cuban environmental law: post-1959 "Cuban law promoted public and collective values from the start. Environmental laws fit easily into this framework" (Ibid.). Second, under the state-dominated, centrally planned economy, the government can mobilize national resources and direct national strategy without negotiating with, or seeking to incentivize, private profit-seeking interests. The state has the institutions and procedures needed to mobilize public and sectoral participation. Third, Cuba's world-leading record of anticipating and responding to risk and natural disasters, demonstrated frequently in its response to hurricanes and, more exceptionally, by its containment of the COVID-19 pandemic.[2] Fourth is Cuba's capacity to collect and analyze local data, a precondition for evaluating specific regional threats.[3] Fifth, the contribution of geographers and environmentalists, who introduced attention to the natural environment into the revolutionary leadership post-1959. Outstanding among them is Antonio Núñez Jiménez, founder of the Cuban Speleological Society in 1940 at the age of 17. A socialist and professor of geography in the 1950s and author of *Geography of Cuba* (published in 1954), he served as a captain in Che Guevara's Rebel Army column and from 1959 headed the National Institute of Agrarian Reform and the National Commission for the Cuban Academy of Sciences, among other roles (Funes Monzote 2019a). President Fidel Castro, who was inspired by Núñez Jiménez, also propelled Cuba's environmental movement, particularly from 1992. "Fidel was the main driving force for the incorporation of environmental concerns into the Cuban policy," affirmed Tirso W. Sáenz (2020), who headed Cuba's first environmental Commission.[4] Sixth, the institutional support provided by the Cuban Communist Party (Partido Comunista de Cuba, or PCC) whose "open endorsement of environmental protection and sustainable growth provides significant legitimacy to environmental programs," according to Houck (2020, 11).

Before discussing *Tarea Vida*, this chapter will review the earlier, incremental focus on environmental protections and Cuba's achievement in terms of sustainable development. It also highlights how the United States' 60-year blockade of Cuba obstructs access to the external financing required to make costly environmental investments. The blockade is the principal obstacle to Cuban development; constraining the island's trading relations, impeding access to foreign investment and technology, and limiting the possibilities for the development of the national economy.

THE CUBAN REVOLUTION COMMITS TO ENVIRONMENTAL PROTECTION

The first of Cuba's "national parks" was designated in 1930. It was mostly a paper exercise until the revolutionary leadership's Agrarian Reform Law in 1959, which added new areas to a program of managed reserves. Chapter 1 of the Law, "On the Conservation of Forests and Soils," also gave the state responsibility for protecting natural areas, initiated reforestation programs, and excluded forest reserves from distribution to agricultural collectives (Houck 2000).

Meanwhile, there was a move to uproot sugar cane fields which were associated with a long history of colonialism, slavery, racism, imperialism, trade dependence, endemic unemployment, exploitation, and the structural weaknesses in the economy. This was soon reversed, however, as Cuban trade shifted from the United States to the USSR and the Socialist Bloc countries. In the socialist camp's international division of labor, Cuba was to continue as a major producer and exporter of sugar, importing food, fuel, manufactures, and industrial goods from the socialist countries. Cuban exports were to provide the basis for state accumulation, which would then facilitate investments for transforming the structure of the economy. Thus, in the 1960s through to the 1980s, Cuba continued to be characterized by increasingly industrialized agricultural production. The government adopted the "Green Revolution" of mechanized agriculture on its predominantly state farms, relying on imported natural gas-based fertilizers, oil-based pesticides, and diesel-powered machinery. The island used 90,000 tractors (Arthur Morgan

Institute for Community Solutions 2006). Yields were high, but production was directed toward exports (sugar, tobacco, and citrus) and the island was unable to feed its own population.

Cuban environmental historian, Reinaldo Funes Monzote, explained that, as in other socialist countries, the Cuban leadership believed "that in the socialist social regime there were no antagonistic contradictions between economic activities and the ecosystems, due to the end of the exploitation of man by man and the preposition of collective interests over private benefits" (Funes Monzote 2019b, 408). Science and technology were to be harnessed in the battle by man to dominate nature (Funes Monzote 2019a). However, from the early 1970s, as the world was waking up to the limits of growth presented by environmental factors, this view was challenged in Cuba.

In 1970, an Earth Day event took place in the United States. In 1971, UNESCO launched the intergovernmental scientific program Man and the Biosphere (MAB). In 1972, the United Nations hosted the Conference on the Human Environment, its first major event on environmental issues, in Stockholm, Sweden (Funes Monzote 2019b). Cuba prepared a report for the conference, although subsequently the Cuban delegates did not attend, in protest of the exclusion of the German Democratic Republic. The following year, 1973, a U.S. specialist in the management of national parks and protected areas, Kenton Miller, visited Cuba to study the island's national parks and wildlife; his report helped generate consensus about the need to consider the environment, as well as "man," in development and provided guidance on planning a conservation program.

In 1975, the First Congress of the PCC approved a "scientific policy thesis," which indicated the "need to increasingly pay attention to the protection and improvement of the environment and the exploitation of natural resources" (Funes Monzote 2019b, 411). There was a call to prioritize scientific technical investigations into erosion and salinization of soils, groundwater contamination, exploitation of mineral reserves, and the impact on flora and fauna, as well as to set up a national body to recommend legislative measures and new technologies. In 1976, a new Constitution was approved. Article 27 stated: "To assure the wellbeing of its citizens, the state and the society will protect nature. It is incumbent on the responsible agencies and on each citizen as well to see that the waters and air are maintained in a clean condition and that the soil, flora and fauna are protected." Thus, according to the UN's Food and Agricultural Organization (FAO), "Cuba was one of the first countries to include environmental issues in its constitution (Article 27 of 1976) and issued environmental laws well before the Bruntdland Report [1987] presented its principles for sustainable development" (FAO 2021). The Brundtland Report, *Our Common Future*, was published in 1987 by the World Commission on Environment and Development, established by the

UN in 1983. It defined sustainable development as "development that meets the needs of the present without compromising the ability of future generations to meet their own needs."

In 1976, a National Commission for the Protection of the Environment and the Rational Use of Natural Resources (Comisión Nacional de Protección del Medio Ambiente y los Recursos Naturales, or COMARNA) was set up in Cuba under a new State Committee for Science and Technology. It combined all the agencies with environmental responsibilities, including those in the area of public health, water resources, fisheries, and sugar. COMARNA's key task was to "establish a system based on the most advanced scientific-technical knowledge to allow the harmonious development of human activity with the least possible degree of environmental and natural resource damage" (Funes Monzote 2019b, 412). Sáenz, then-first vice president of Cuba's Academy of Sciences, was named Head of COMARNA, which had a nucleus of half-a-dozen technical specialists and representatives in each province. "COMARNA did not have executive powers; we could not impose fines or stop contaminating practices" he explained (Sáenz 2020).[5] A Scientific Commission for the Conservation of Flora and Fauna was set up to cooperate with COMARNA, addressing the degradation of forests, water contamination, hunting regulations, unnecessary clearings, and other activities.

Sáenz visited several Council of Manual Economic Assistance (known by the acronyms CMEA or COMECON) countries to learn about their legislation on the environment. "They were excellent," he said. "The problem was how to fight internally between the focus on good environmental practices versus the interest in producing profits; many times, the latter won. On the other hand, I saw good instruments like the hydrometeorology system in the Soviet Union; and the water sewage system in Hungary" (Sáenz 2020).

In 1980, Raúl Castro, then minister of the Armed Forces, proposed a law in the National Assembly of People's Power to protect the countryside from the impact of nickel extraction in Nicaro, Moa, and other northeastern zones. In the subsequent discussion, it was agreed to broaden the scope of the proposed law. The result was Law 33 for the Protection of the Environment and the Rational Use of Natural Resources, enacted in 1981. It was accompanied by complementary laws concerning the land, preservation of human health, defense of fishery resources, animal and plant health, seed certification and preservation of terrestrial waters, protection of hydraulic works and water sources, the exploitation and preservation of mining resources, tourist resources and the landscape, protection of the atmosphere, and the rules for beekeeping and honey (Funes Monzote 2019b). This legal focus was accompanied by an effort to raise the level of knowledge and commitment among the population by addressing environmental issues in the media.

Studies were published between 1981 and 1982 about the contamination of water, particularly the Bay of Havana, and the Bay of Cienfuegos.[6] Meanwhile, reforestation projects focusing on the mining zones were bolstered. By that stage, the number of trees planted in Cuba had increased from 10 million in 1959 to 550 million (Funes Monzote 2019b, 421). In 1983, COMARNA produced a detailed report of the country's environmental situation, with a breakdown for each province. Nonetheless, Funes Monzote concluded "The stress was still, somehow, to see the environmental issue as something subordinate or parallel to economic development" (Funes Monzote 2019b, 424). In 1986, *Plan Turquino* was launched; a socio-economic program to stabilize mountain communities and increase the independence of mountain areas; followed by *Plan Manatí*, a program to preserve the balance among agricultural areas, forests, and watersheds (FAO 2021). In the meantime, UNESCO's program Man and the Biosphere recognized four Cuban sites in the 1980s: the Sierra del Rosario Biosphere Reserve, Guanahacabibes Peninsula, Cuchillas del Toa, and Baconao.

According to Sáenz, "the work done by the State Committee and particularly by COMARNA was considered poor (not officially) and it was dissolved (along with other State Commissions that were part of a gigantic, inefficient, and bureaucratic superstructure). Its functions were assigned to the ACC [Cuban Academy of Sciences]" (Sáenz 2020). Framing these weaknesses in terms of legal enforcement mechanisms, Houck explained that "In 1990, Cuba made one more try at collective management in Decree Law 118, allocating specific environmental responsibilities among more than eight separate Ministries and authorizing a ninth, a new Ministry of Science, Technology and Environment (CITMA) to replace COMARNA" (Houck 2000, 15). The process was interrupted by the existential and severe economic crisis that Cuba confronted with the collapse of the Socialist Bloc between 1989 and 1991.

THE PIVOT TO SUSTAINABLE DEVELOPMENT IN THE 1990S

With the transition to capitalism in Eastern Europe and the USSR, Cuba lost around 86 percent of its trade and investment, sending the economy plummeting some 35 percent in three years and causing severe shortages in every sector (Rodríguez 2013). During the resulting years of austerity and scarcity, known as the "Special Period," the island was compelled to transform its highly mechanized, chemical-based agricultural production; state farms were split into cooperatives operating organic and eco-agricultural techniques. Cuban research centers had begun studying sustainable agriculture, including biological pest controls, before the crisis in response to the effects of chemical fertilizers on land and coastal water quality. Already by the late 1980s, ninety sugar cooperatives were converted for use as organic fertilizer, and between 1989 and 1992, there were sharp reductions in the use of chemical fertilizers and herbicides (Houck 2000, 63). The state quickly channeled this work into replacing the chemical inputs no longer available with locally produced biopesticides and biofertilizers (Álvarez et al. 2006). In addition, they turned to earthworms, compost, other organic fertilizers, animal and green manures, and the integration of animal grazing. In 1996, the National Association of Small Farmers (Asociación Nacional de Agricultores Pequeños, or ANAP) added "sustainability" to its goals and promoted agro-ecological farming techniques nationwide. Combining scientific expertise and popular engagement, Cuba achieved extraordinary successes in the shift to organic and urban farming, city gardens, local food production, as well as mass cycling and collective hitchhiking for transportation.

Against this backdrop, Fidel Castro went to Rio de Janeiro, Brazil, for the UN's 1992 Conference on Environment and Development, or "Earth Summit." The Cuban president delivered an uncharacteristically short and appropriately alarming speech asserting that the exploitative and unequal international relations resulting from colonialism and imperialism underlay the rapacious environmental destruction fueled by capitalist consumer societies. These impacts were most felt by underdeveloped countries and the survival of humanity was under threat as a result of environmental devastation, he stated. Houck noted that while in Rio, Fidel Castro met with grassroots activists and oppositional environmental organizations, so that "the revolutionary aspect of this movement was too obvious to ignore. It must have struck a chord. Castro embraced this environmental revolution, brought it home, and put it in the new constitution, unleashing a chain reaction that is still running its course" (Houck 2000, 18).

Subsequently, Article 27 of the Constitution was strengthened to say:

> The State protects the environment and the natural resources of the country. It recognizes their close link with the sustainable economic and social development for making human life more sensible, and for ensuring the survival, welfare, and security of present and future generations. It corresponds to the competent organs to implement this policy. It is the duty of citizens to contribute to the protection of the water and the atmosphere, and to the conservation of the soil, flora, fauna, and all the rich potential of nature. (Constitution of the Republic of Cuba 1992)

The Cuban economy had to be restructured for reintegration into the capitalist world market but in the process of recovery, environmental protections were to be prioritized and accompanied by a commitment to sustainable development, to redress past environmental harm and minimize future degradation of air, water, and land resources (Whittle and Santos 2006).

The new Ministry of Science, Technology, and Environment (Ministerio de Ciencias, Technología y Medio Ambiente, or CITMA), created in 1994, was tasked with devising the National Environmental Strategy, which was then adopted in 1997. That same year, Law 81, Law on the Environment, was approved unanimously in the National Assembly, giving CITMA broad powers in the evaluation of environmental impact and the issuing of environmental licenses (Houck 2000). "This law is regarded as more ambitious in its goals and details than any comparable legislation in the United States or Western Europe," stated Funes Monzote (2019b, 476).

Also in 1997, the National Council of Basins was set up to administer programs for the integral management of hydrographic basins. Other programs initiated in this period concerned desertification and drought, climate change, and biodiversity. Mountainous regions received special attention through the *Turquino-Manatí* Plan and the establishment of forest farms. In 1995, a Center for National Protected Areas was established which led to the 1999 Law of Protected Areas. Several areas in Cuba became UNESCO World Heritage Sites, biosphere reserves, and wetland sites designated as being of international importance under the Ramsar Convention (an international treaty for the conservation and sustainable use of wetlands). By the early 2000s, Cuba's system of national protected areas covered 22 percent of the national territory, concentrated in marine zones. This period also saw improvements in river recoveries (Funes Monzote 2019b).

While the economically necessary expansion, with foreign capital, of mining and tourism heightened environmental tensions, organic farming and sustainable agriculture were embedded into Cuban development. From 1994, the Urban Agriculture Department promoted the local production of organic foods for Havana under the slogan "food production in the neighborhood, by the neighborhood and for the neighborhood." This was extended nationwide through the National Urban Agriculture Group, established in 1998. The use of agrochemicals in urban gardens was prohibited by law. The production of vegetables and fresh herbs jumped a thousand-fold from 4,000 tons in 1994 to 4.2 million in 2005 (Koont 2009, 47). Produce was sold directly to consumers in local stands and farmers' markets.

Meanwhile, and despite the economic crisis, Cuba's human development indicators continued to improve. Thus, in 2006 the World Wildlife Fund's (WWF) Living Planet Report identified Cuba as the only country achieving sustainable development; improving the quality of human life while living within the carrying capacity of its ecosystem (WWF 2006). In 2015, the Global Footprint Network identified Cuba as the only country attaining "very high human development" at a cost of less than 1.7 global hectares per capita (Global Footprint Network 2015).[7] In 2019, Jason Hickel's new Sustainable Development Index (SDI), which updates the UN's Human Development Index (HDI) by measuring the ecological efficiency of human development, ranked Cuba best in the world between 1990 and 2015 (Hickel 2020). Notwithstanding Cuba's international leadership in sustainable development, the island is not exempt from the threat of climate change.

Funes Monzote concludes that Cuba "is not an environmental paradise, and there are many limitations and insufficiencies. In some cases, they have to do with bad decisions or voluntarism. In others, with economic restrictions and sanctions, which, while reducing the pressure on the ecosystems, make it impossible or difficult to solve many problems that depend on the availability of financial resources" (Funes Monzote 2019b, 478). However, sustainable development and environmental protection is embedded into Cuba's development strategy. "Natural resources and the environment" is one of six strategic areas in the "National Plan of Economic and Social Development to 2030" which was launched in 2017.[8] The "general objectives" are: "(1) Guarantee the protection and rational use of natural resources, the conservation of ecosystems, and care of the environment and the natural heritage of the nation for the benefit of society; (2) Raise the environmental quality; and (3) Reduce the country's vulnerability to the effects of climate change through the gradual execution of the State Plan [*Tarea Vida*] for confronting it." It is to *Tarea Vida* that this chapter now turns.

THE CHALLENGE OF CLIMATE CHANGE: ANTECEDENTS TO *TAREA VIDA*

The first domestic study to assess the impact of climate change on Cuba was carried out in 1991, prior to the USSR's collapse and the island's shift to sustainable agriculture. Following two devastating hurricanes that hit Cuba in 2004, national investigations into dangers, vulnerabilities, and risks were increased. In 2005, the National Defense Council approved a directive for the planning, organization, and preparation for disaster situations. In 2006, studies in dangers, vulnerability, and risk were initiated and in 2007 the Council of Ministers approved the first program to confront climate change. One of the tasks, delegated to CITMA, was to carry out a major investigation into coastal hazards and vulnerabilities, projecting forward to 2050–2100 (Pell del Río and Herrera Cruz 2017; Tamayo León 2017).

Known as the "macro-project," this incorporated ongoing annual investigations carried out by multidisciplinary groups of scientists. In February and March each year, the principal results are presented to the country's leadership to inform policy. Minister Elba Rosa Pérez Montoya, the head of

CITMA since 2012, described it as a "fundamental tool that will continue, and that involves various institutions and many experts; that advances and consolidates as scientific research of practical utility" (Tamayo León 2017). Forecasts are based on physical-mathematical models. "In addition, the results produced are one hundred percent Cuban [. . .] and have an impressive coherence with the studies of the Intergovernmental Panel on Climate Change (IPCC) of the United Nations" she added (Ibid.).

In 2011, the Council of Ministers approved six directives concerning climate change. In 2015, CITMA updated the documents previously approved by the Council of Ministers for confronting climate change and was commissioned to present a state plan on the issue, to convert the scientific results into a higher ranking policy. Two years later, on April 25, 2017, *Tarea Vida: State Plan to Confront Climate Change* was approved at the highest level of government. The state plan draws directly on the macro-project that began in 2007, as well as upon existing apparatuses in Cuba, including: INSMET; the Institute for Hydraulic Resources; the Civil Defense (Defensa Civil, or DC) system; thematic networks (hydraulic, meteorological, seismological, geodesic, oceanographic, radiologic, and sanitary); the National Atlas of Cuba; and directives of the Council of National Defense for the Reduction of Disasters (Pell del Río and Herrera Cruz 2017).

CLIMATE CHANGE: AN EXISTENTIAL THREAT

Rising sea levels, coastal erosion, extreme weather events, and freshwater shortages resulting from reduced rainfall and the salinization of aquifers are all climate change impacts that will increasingly assault Cuba. According to Cuban scholars, Silvia Miriam Pell del Río and Juan N. Herrera Cruz, "Coastal flooding caused by extreme weather events represents the greatest danger and damage to the natural and built heritage" (Pell del Río and Herrera Cruz 2017). By 2017, 82 percent of the 499 coasts that Cuban investigators had evaluated showed signs of accumulated erosion and the deterioration of coastal protections, such as sandy beaches, wetlands, and the ridges of coral reefs (Ibid.). This is particularly serious for Cuba, as with other Caribbean islands, given the country's dependence on income from tourism.

Cuba is experiencing less rainfall, higher temperatures, and increasing drought. Between 1961 and 1990, there was a 0.9° Celsius rise in the average surface temperature, with an increase in the average minimum temperature of more than double and a significant reduction in the diurnal temperature range (the difference between the daily maximum and minimum temperature). Meanwhile, droughts are more frequent and last longer, especially in summer, and while there has been a 10 percent reduction in annual rainfall, winters have seen heavy downpours

(Planos Gutiérrez 2017). Meanwhile, the island was battered by an unprecedented nine intense hurricanes between 2001 and 2017 (Tamayo León 2017). The most probable scenarios for Cuba have the average air temperature rising by between 2.6 and 4.5° Celsius with annual precipitation falling by between 15 percent and 60 percent in respect to current values. The most favorable scenarios see water availability falling by 37 percent below the 1961–1990 baseline: meaning a significant reduction in the island's water potential (Planos Gutiérrez 2017). This will likewise seriously impact on food production.

The sea level has risen by an average of 6.77 centimeters between 1966 and 2017, but at an accelerated rate in the last five years of that period. Pell del Río and Juan N. Herrera Cruz (2017) reported that "future projections indicate that the elevation of the mean sea level could reach up to 27 centimeters by 2050 and up to 85 centimeters by 2100, values that correspond to the estimated ranges for the entire planet; which will imply a slow decrease of our emerged surface and a gradual growth of salinization, since our underground aquifers will have a level of impact to take into account."

Since the 1990s, Cuban scientists have worked on climate change modeling in Cuba and the Caribbean. However, Eduardo O. Planos Gutiérrez, university professor and researcher at INSMET, notes the level of uncertainty in models, especially regional models mapping longer-term prospects. Climate scenarios are not forecasts, he points out, but plausible representations of a future that depends on the evolution of the world's socioeconomic model, especially in terms of emissions and the absorption of greenhouse gases associated with that. Cuban scientists are working with the Center for Climate Change for the Caribbean Community (CCCCC) in collaboration with the Hadley Center in the United Kingdom, which provides climate science and services, to develop and employ the PRECIS Regional Model, which allows them to represent the future climate of countries as small as those in the Caribbean (Planos Gutiérrez 2017).

However, Cuba's climate modeling does not include the predicted rise in the sea water level, points out Plano Gutiérrez, adding that: "Climate and water are two essential resources for life that should not be separated" (Ibid.). With this impact factored in, he concludes that the Cuban insular platform, and the way the seas adjacent to Cuba are related to the ocean, will suffer significant changes, including gradual modification of the physical geographic, hydrographic, and hydroclimatic characteristics of the insular platform; Considerable reduction in the low areas of the island of Cuba, of the keys and the Ciénaga de Zapata, together with the disappearance of numerous Keys with heights under 0.5 meters; Increased tidal fluctuations and non-periodic variations in sea level, which will increase during severe atmospheric events; Coastal erosion, by up to 7 kilometers; Increase in the speed of marine currents; Increase in the depth of the platform; Increase in the exchange between the ocean and the adjacent seas; and an alteration in the spatial distribution of sediments, as a consequence of changes in the erosion accumulation process along the coast.

This coastal marine scenario, compounded by the reduction of rainfall, will intensify the deficit of potential freshwater availability, as a consequence of marine intrusion on coastal aquifers (Ibid.).[9] According to Planos Gutiérrez, assuming the sea level rises to 85 centimeters, with a 25 percent reduction in precipitation in the wet period, the probable scenario by 2100 is that saline intrusion will have advanced 4–5 kilometers in the horizontal direction and risen 19–21 meters in the vertical direction. He concludes, "This situation would imply a significant reduction in groundwater delivery and, in the less powerful coastal aquifers, it could represent the cessation of pumping, due to the definitive salinization of its reserves" (Ibid.).

The prospects make for somber reading. Cuba's hydraulic resources sector is already under great stress as a result of climate change, and it will be among the most severely affected areas in the future, having a major impact on other resources and sectors. Tensions will be heightened over water management with competing demands; on the one hand, to increase national food production, as part of the policy to substitute imports, and, on the other hand, to sustain the equilibrium of the hydraulic system, protecting the ecosystem and biological diversity. Agricultural production will be affected by the increasing frequency and intensity of droughts, an increasingly arid climate, the decrease in the real evapotranspiration of ecosystems and a pronounced water deficit.[10] The repercussions are diverse: falling agricultural yields; certain crops disappearing, including rice, tobacco, and pork; vectors and plagues increasing; excessive heat generating suffocating atmospheres in houses designed for lower temperatures; forced relocation of homes as sea levels rise; and the increased risk of hydrometeorological disasters.

Planos Gutiérrez (2017) warns that Cuba's future climate could be, "more arid and extreme, characterized by prolonged and frequent droughts and severe water deficits. The dry landscapes of the Eastern zone will intensify, and this will progressively advance toward the Western zone; producing a transformation from humid to dry sub-humid tropical climate, with the threat of desertification." Arguably, only the kind of sweeping state mobilization presented by *Tarea Vida* could stand a chance of confronting such a future.

TAREA VIDA: THE STATE PLAN TO CONFRONT CLIMATE CHANGE

Tarea Vida involves progressive implementation in stages that are split into short-term (to 2020), medium term (to 2030), long term (2050), and very long term (2100). By 2017, 103 studies had been concluded throughout the country on the danger, vulnerability, and risk presented by climate change. In the first stage, studies related to hydrometeorological dangers (strong winds, flooding due to sea penetration, and heavy rain) were completed. The second stage involved studies of drought, fires in rural areas, and dangers of a geological, technological, or sanitary nature. The results of these investigations informed subsequent measures of prevention, preparation, response, and recuperation, to reduce risks and vulnerabilities (Pell del Río and Herrera Cruz 2017).

The cost of relocating industries is prohibitively high for the Cuban state, so the emphasis is on adapting productive entities to the consequences of climate change. This requires giving attention to the human settlements, which have built up around industrial and tourist zones. A group of priority areas was selected based on priorities in the following order: first, the preservation of lives; second, where care for the natural space was necessary to guarantee physical and food security; and third, zones with tourist developments (Pell del Río and Herrera Cruz 2017; Tamayo León 2017). The priority areas are the southern part of both Artemisa and Mayabeque provinces; the north coast of Havana and Havana Bay; Mariel special development zone; Varadero and its tourist corridors; the tourist Keys of Villa Clara and the North of Ciego de Ávila and Camagüey; the north and south coast of Ciego de Ávila; the north coast of Holguín; the Bay of Santiago de Cuba; and coastal cities threatened by the gradual rise of the sea (Cienfuegos, Manzanillo, Moa, Niquero, and Baracoa). Priority areas also include coastal settlements not included in the previous zones: areas expected to disappear between the years 2050 and 2100 (Sancti Spíritus, Camagüey, Pinar del Río, and Villa Clara); sandy beaches with intense erosion not included in the previous areas that would disappear if affected by extreme weather events; other areas of tourist and recreational interest (Camagüey, Pinar del Río, Granma, Holguín, Las Tunas, and Isla de la Juventud); and unprotected coastal areas with saline intrusion (Pinar del Río, Matanzas, Granma, Camagüey, Cienfuegos, and Sancti Spíritus). This list demonstrates that, as Minister Pérez Montoya said, "Cuba is completely vulnerable." But she added that "we cannot take action in all places at the same time. We must go progressively" (Tamayo León 2017). It is for that reason *Tarea Vida* is broken into stages, actions, and tasks.

TAREA VIDA: STRATEGIC ACTIONS AND TASKS

The high cost of environmental investments means that economically blockaded Cuba has to carefully plan *Tarea Vida* according to the set of "strategic actions and tasks" listed below:

Strategic Actions

1) Do not allow the construction of new homes in threatened coastal settlements that are predicted to disappear as a result of permanent flooding. Reduce population density in low-lying coastal areas.[11]
2) Develop constructive concepts in infrastructure, adapted to coastal flooding in low-lying areas.
3) Adapt agricultural activities, particularly those with the greatest impact on the country's food security, to changes in land use due to rising sea levels and drought.
4) Reduce the cultivation in areas near the coasts or affected by saline intrusion. Diversify crops, improve soil conditions, introduce and develop varieties resistant to the new temperature scenario.
5) Plan, within specified timescales, the urban reorganization processes of the threatened settlements and infrastructures, in correspondence with the economic conditions of the country. Start with lower-cost measures, such as induced natural solutions (beach recovery, coastal reforestation, and so on).

Tasks

1) Identify and undertake actions and projects (coastal protection of cities, relocation of human settlements, integral recovery of beaches, mangroves and other protective natural ecosystems, hydraulic works and coastal engineering, among others) of adaptation to climate change, of a comprehensive and progressive character, necessary to reduce the existing vulnerability in fifteen zones identified as priorities. These are assessed in the order of threatening population, food security, and the development of tourism.
2) Implement the legal norms necessary to support the execution of the state plan, as well as ensure its strict compliance, with particular attention to measures aimed at reducing vulnerability of the built heritage, prioritizing threatened coastal settlements.
3) Conserve, maintain, and fully recover the sandy beaches of the Cuban archipelago, prioritizing urbanized beaches for tourist use and reducing the structural vulnerability of the built heritage.
4) Ensure the availability and efficient use of water to combat drought, based on the application of technologies for saving water and satisfying local demands. Increase hydraulic infrastructure and its maintenance, as well as the introduction of actions to measure efficiency and water productivity.
5) Direct reforestation toward the maximum protection of the soils and waters in quantity and quality, as well as the recovery of the most

affected mangroves. Prioritize reservoirs, canals, and hydroregulatory strips of the tributary basins of the main bays and the coasts of the insular platform.

6) Stop the deterioration of, as well as rehabilitate and conserve, the coral reefs throughout the archipelago, with priority given to the ridges that line the insular platform and protect urbanized beaches for tourist use. Avoid overfishing of fish that favor coral reefs.
7) Maintain and introduce the scientific results of the macro-project on hazards and vulnerability in the coastal zones (2050–2100) within the land and urban plans. In addition, include the studies regarding disaster vulnerability and risk reduction. Use this information for future decision-making.
8) Implement and control the measures of adaptation and mitigation of climate change derived from the sectoral policies in the programs, plans, and projects linked to food security, renewable energies, energy efficiency, land and urban planning, fishing, agricultural activity, health, tourism, construction, transport, industry, and forest management.
9) Strengthen monitoring, surveillance, and early warning systems to systematically assess the status and quality of the coastal zone, of water, drought, forest, and human, animal, and plant health.
10) Prioritize measures and actions to increase the perception of risk, improve the knowledge and participation of the entire population in tackling climate change, and promote a culture of water conservation.
11) Manage and use the international financial resources available, from global and regional climate funds, as well as from bilateral sources, to execute the investments, projects, and actions that derive from each of the state plan's tasks.

NATIONAL MOBILIZATION AND COMMUNITY PARTICIPATION

Cuban political economist, Alfredo Carballo, describes *Tarea Vida* as a "cultural change" (Carballo 2021). To advance this, and in line with Task 10, a communication strategy has been developed to increase citizen awareness of the dangers of climate change, inform the population about *Tarea Vida* in general terms and advise communities about the specific problems they face, and the measures required in each case. This task mobilizes many specialists, leaders of the provinces, the government and the PCC, representatives of entities of the state's central administration, DC, physical planning, and territorial Risk Reduction Management Centers (RRMCs). These are established institutions and apparatuses of Cuba's socialist state, with long traditions of mobilization at the community level.

In the months following the approval of *Tarea Vida*, minister Pérez Montoya toured the country, with colleagues from CITMA, DC, and other entities, giving presentations about *Tarea Vida* to provincial and municipal authorities about the risks faced and the action necessary to confront them. They showed video footage evidencing, for example, the erosion of beaches and the vulnerability of some communities. The footage includes material of Fidel Castro addressing the issue, harnessing his "moral authority" to win consensus. "The meetings have generated a lot of attention and greater responsibility," the minister told the daily newspaper *Juventud Rebelde* during the tour. "However, I think that more needs to be done, that is why the State Plan has a Task related to public perception [. . .] we are developing a strategy aimed at the media, to keep them on board [. . .] we must give systematic and detailed explanations" (Tamayo León 2017).

The minister acknowledged concern, and in some cases resistance, from individuals living in the most vulnerable settlements where new constructions have been prohibited, and there are plans to decrease the population. Many people have a strong affinity with the location where their parents and grandparents lived before them. Pérez Montoya told *Juventud Rebelde*:

> We must explain that the coastline on sandy beaches is receding 1.2 meters on average each year, that marine intrusion into underground basins and the salinization of soils is increasing, affecting agricultural production. . . . We must continue explaining, arguing, persuading, especially in communities near the sea. It is also necessary for scientists to continue working on the development of crop varieties that are more resistant to climate change, in order to plant in these conditions, in those places. After explaining *Tarea Vida*, there is a positive reaction, but we must continue to insist on the risks, on the possible negative impacts of climate change. (Tamayo León 2017)

CITMA provided provincial authorities with informational material, advising them to go into the communities, especially the most remote ones, to explain *Tarea Vida* and organize neighborhood debates.

> It is necessary to go to each town, to each settlement [. . .] to all the vulnerable communities and those that appear less at risk, so that the population knows the effects of climate change; because it is not only that the sea is going to penetrate, but as consequence of the increase in temperature the risk of contracting new and re-emerging diseases produced mostly by vectors will also increase. There are a number of consequences and it is very important to know them. (Tamayo León 2017)

This communication strategy is reinforced more broadly by coverage of climate change and food sustainability issues in both the education system and the media.

CONTROLLING IMPLEMENTATION

In line with Task 2, the establishment of legal norms for ensuring the implementation of *Tarea Vida*, CITMA has proposed three measures. First, a modification of Decree Law No. 212 (instituted in 2000), on coastal zone management, accompanied by other regulations. Second, a new Decree Law specifically for tackling climate change and which will include a group of economic, tax, and financial instruments. And the third, the need to update Environmental Law 81, approved in 1997 (Ibid.).

A comprehensive environmental inspection and control system is in place to oversee the execution of *Tarea Vida*.[12] Each task has been allocated a stage. In each of Cuba's fifteen provinces (and one special municipality, the Island of Youth), CITMA directs the tasks to be prioritized and the timescale for completion. "If we give up to a certain date for the recovery of sandy beaches, or a stretch of the coast, they must comply," explains Pérez Montoya. Because of the high investments required for environmental engineering, the economic resources necessary to complete the plans will be made available on a yearly basis by CITMA and the Ministry of the Economy and Planning (Ministerio de la Economía y Planificación, or MEP). For example, the Malecón (the broad esplanade, roadway, and seawall that stretches for 8 kilometers along the coast in Havana), which has been flooding more frequently, is included among the fifteen vulnerable areas initially prioritized in Task 1. However, its renovation requires a multimillion-dollar investment which is not available, so the work has been divided into installments according to what the country can afford at each stage. A similarly detailed analysis has been carried out for the other fourteen priority spots, with specified timelines (Tamayo León 2017).

CITMA is responsible for demanding the planning and execution of the investment processes prioritized in *Tarea Vida* by the institutions of the central state administration, the higher state entities, national enterprises, and the Councils of the Provincial Administration. These bodies are instructed to improve their search for the external financing that the Cuba government needs to carry out *Tarea Vida*, applying to the Global Environmental Facility, the Green Climate Fund, and other bilateral funds. International climate finance is emerging as one of the very few sources of development assistance that the United States is not able to block. "The State Plan has to use bilateral collaboration and what we can achieve through these international organizations," the minister explained (Tamayo León 2017).[13]

The U.S. blockade of Cuba makes the challenge of securing external financing more complex, as the U.S. Treasury works to block financial flows to the island, not just from the United States but from third-party countries and organizations internationally. As in so many areas of Cuban development, progress is mired by the significant challenges presented by the U.S. blockade, which obstructs Cuba's engagement with the rest of world.

TOMORROW WILL BE TOO LATE

The objective of *Tarea Vida* is to protect Cuba from the disastrous effects of climate change: to safeguard the population, their physical and food security, as well as the country's built heritage and the development of tourism. The potential for the plan's success rests on the combination of political and resource commitment by the Cuban state, the scientific-technical expertise at its core, the institutional apparatuses at its disposal, and the growing level of popular awareness and participation.

Cuba is ready to match its world-leading performance in human development indicators and sustainable development with an exemplary response to climate change. However, the government requires financial resources that must be provided externally, which means navigating the U.S. blockade. If the Cuban government succeeds in accessing sufficient finance to harness its skills and capabilities, its state commitment, its system of economic management and participation, *Tarea Vida* will provide a lesson for other developing countries, particularly small island nations facing similar threats. In the words with which Fidel Castro concluded his speech in Rio, 1992: "Tomorrow will be too late for what we should have done a long time ago."

NOTES

1. Forest cover was 95 percent prior to Cuba's colonization; down to 52 percent in 1900 and between 8–14 percent by 1959 (Funes Monzote 2009).
2. In January 2021, Cuba's Covid-19 death rate was 13 per million, compared with over 1,000 per million in the US (DaniFilms and Helen Yaffe 2020).
3. Thanks to Dr. Emily Morris for pointing out this key factor.
4. All translations from the original Spanish are the author's own.
5. In the 1960s, Tirso W. Sáenz served as Vice Minister of Scientific Development in the Ministry of Industries under Ernesto "Che" Guevara.

6. Clearly, progress was limited as in the mid-1990s the UN described the Havana Bay as one of the ten most polluted in the world (Houck 2000).

7. Seven other countries qualified with "high human development." By comparison, the average person in Britain consumes around eight global hectares (Global Footprint Network 2015).

8. The other five strategic areas are: effective, socialist government and social integration; productive transformation and international insertion; infrastructure; human potential—science, technology and innovation; human development, social justice and equity.

9. Planos Gutiérrez draws on the IPCC estimates for the rise in sea levels (27 centimeters by 2050 and 85 centimeters by 2100) and on localized Cuban studies and the results of the macro-project "Hazard and vulnerability scenarios for Cuba coastal zone, associated with the rise of the mean sea level for the years 2050 and 2100" (Planos Gutiérrez 2017).

10. Evapotranspiration is the sum of water evaporation and transpiration from a surface area to the atmosphere.

11. Decree Law 212 of 2000 had already prohibited most constructions, new developments, waste disposal and other activities on Cuban coasts.

12. This work is carried out by institutions of the Central State Administration, the Provincial Administration Council and the Municipal Administration Council.

13. In June 2020, the UN's FAO signed an agreement to contribute $38.2 million to a joint project with the Cuban government to foster climate change resilient agricultural food production, benefiting 240,000 Cubans while reducing greenhouse gases (*Cubadebate* 2020).

BIBLIOGRAPHY

Álvarez, Mavis, Martin Bourque, Fernando Funes, Lucy Martin, Armando Nova and Peter Rosset. 2006. "Surviving the Crisis in Cuba: The Second Agrarian Reform and Sustainable Agriculture." In *Promised Land: Competing Visions of Agrarian Reform*, edited by Peter Rosset, Raj Patel and Michael Courville, 225–248. Oakland, CA: Food First Books.

Arthur Morgan Institute for Community Solutions. 2006. *The Power of Community: How Cuba Survived Peak Oil*. Documentary.

Carballo, Alfredo. 2021. Interview with Helen Yaffe. January 21. Havana.

Constitution of the Republic of Cuba. 1992. Accessed February 24, 2020. https://www.cubanet.org/htdocs/ref/dis/const_92_e.htm.

Cubadebate. 2020. "Firman Cuba y la FAO acuerdo para impulsar sistemas agroalimentarios más sostenibles." June 1. Accessed July 14, 2020: http://www.cubadebate.cu/noticias/2020/06/01/firman-cuba-y-la-fao-acuerdo-para-impulsar-sistemas-agroalimentarios-mas-sostenibles/.

DaniFilms and Helen Yaffe, producers. 2020. *Cuba & Covid-19: Public Health, Science and Solidarity*. November. Accessed December 2020. https://www.youtube.com/watch?v=lGYHwldJ_gY&t=5s.

FAO. 2021. "Cuba." Food and Agricultural Organization (FAO) of the United Nations. Accessed January 20, 2021. http://www.fao.org/forestry/watershedmanagementandmountains/74916/en/.

Funes Monzote, Reinaldo. 2009. *From Rainforest to Cane Field in Cuba: An Environmental History Since 1492*. Chapel Hill, NC: University of North Carolina Press.

Funes Monzote, Reinaldo. 2019a. "*Geotransformación*: Geography and Revolution in Cuba from the 1950s to the 1960s." In *The Revolution From Within: Cuba, 1959–1980*, edited by Michael J. Bustamente and Jennifer L. Lambe, 124–128. Durham, NC: Duke University Press.

Funes Monzote, Reinaldo. 2019b. *Nuestro Viaje a la Luna: la idea de la transformación de la naturaleza en Cuba durante la Guerra Fría*. Havana: Fondo Editorial Casa de Las Américas.

Global Footprint Network. 2015. *Only eight countries meet two key conditions for sustainable development as United Nations adopts Sustainable Development* Goals. September 23. Accessed January 12, 2021. https://www.footprintnetwork.org/2015/09/23/eight-countries-meet-two-key-conditions-sustainable-development-united-nations-adopts-sustainable-development-goals.

Hickel, Jason. 2020. "The Sustainable Development Index: Measuring the Ecological Efficiency of Human Development in the Antropocene." *Ecological Economics* 167: 1–10.

Houck, Oliver A. 2000. "Environmental Law in Cuba." *Journal of Land Use and Environmental Law* 16 (1): 1–81.

Koont, Sinan. 2009. "The Urban Agriculture of Havana." *Monthly Review* 60 (8): 44–63.

Pell del Río, C. Silvia Miriam and Juan N. Herrera Cruz. 2017. "Tarea Vida: plan del estado para el enfrentamiento al cambio climático." *Revista Bimestre Cubana* July-December 3 (47).

Planos Gutiérrez, Eduardo O. 2017. "Sintesis Informativa sobre la segunda evaluación de impactos del cambio climático en cuba y medidas de adaptación." *Revista Bimestre Cubana* July-December 3 (47).

Rodríguez, José Luis. 2013. "Fifty Years of Revolution in the Cuban Economy." In *Cuban Economists on the Cuban Economy*, edited by Al Campbell, 25–61. Gainesville: University Press of Florida.

Sáenz, Tirso W. 2020. Email Interview with Helen Yaffe. June 12.

Tamayo León, René. 2017. "Tarea Vida: Entrevista con Elba Rosa Pérez Montoya, titular del Ministerio de Ciencia, Tecnología y Medio Ambiente." *Juvented Rebelde*. June 10. Accessed March 25, 2020. http://www.juventudrebelde.cu/ciencia-tecnica/2017-06-10/tarea-de-vida.

Whittle, Daniel, and Orlando Rey Santos. 2006. *Protecting Cuba's Environment: Efforts to Design and Implement Effective Environmental Laws and Policies in Cuba*. Pittsburgh, PA: University of Pittsburgh Press.

WWF. 2006. *World Wildlife Fund: Living Planet Report 2006*. Accessed March 26, 2020. https://wwfint.awsassets.panda.org/downloads/lpr_living_planet_report_2006.pdf.

Chapter 10

International Collaboration on the Environment and Marine Conservation in Cuba

Reflections from Environmental Defense Fund

Valerie Miller

For many decades, Cuba has committed itself to an ambitious vision for environmental sustainability and marine conservation. International collaboration between Cuba and the United States on science and the environment has served as a bridge between the two countries, even when no diplomatic relationship existed. This chapter examines Cuba's special role in marine conservation and the importance of collaboration across Cuba, the United States, and the wider region. The chapter will then discuss a selection of important results from a U.S.-based non-governmental organization's twenty years of collaboration with Cuba on marine conservation and the environment, and, in doing so, demonstrate how Cuba is leading the way with results that are having a global impact. Finally, it will analyze the common goals that can guide future environmental collaboration between the United States and Cuba.

The reflections in this chapter stem from Environmental Defense Fund's (EDF) history of working with Cuba since 2000 to protect and sustainably manage marine and coastal resources. Founded in 1967, EDF is a non-governmental organization with a mission to preserve the natural systems on which all life depends. Guided by science and economics, the organization seeks practical, lasting solutions to the world's most serious environmental problems that allow people and nature to prosper together. EDF is a global organization, headquartered in New York that focuses on four programmatic areas: climate and energy, ecosystems, health, and oceans. In its oceans work, EDF aims to make the world's oceans healthier to sustain ecosystems, food sources, and fishing communities, even in the face of climate change. To do so, the organization works around the world with partners to end overfishing and advance sustainable fishing practices and policies built on scientific and economic research. EDF's collaboration with Cuban institutions, which originally focused on marine ecosystem protection, has evolved over time to reflect the priorities of partners and the organization's expertise and now incorporates a broader scope of ongoing sustainability issues.

WHAT MAKES CUBA SPECIAL IN TERMS OF MARINE CONSERVATION?

Cuba is a global biodiversity hotspot and the most important island in the entire western hemisphere in terms of biological diversity (Kritzer et al. 2014). The country boasts many healthy marine ecosystems—mangroves, seagrass beds, and some of the best-preserved coral reefs in the Caribbean (Gonzalez et al. 2018). Cuba is also home to a high percentage of endemic flora and fauna, including 3,200 plants, 15 mammals, 131 reptiles, 65 amphibians, and 26 bird species only found in Cuba (CITMA 2019).

Cuba's commitment to the environment and strong environmental policy is distinct from other countries. In 1997, the Cuban government established the legal framework for a National System of Protected Areas, which resulted in an extensive network of nature reserves, national parks, wildlife refuges, and other protected areas on land and in the sea (Law 81, Article 90). Two years later, the government established a decree law that established eight categories of protected areas and provisions for designating and managing areas (Whittle et al. 2002; Whittle and Rey Santos 2006). In 2021, the network comprises 211 conservation areas, representing 17 percent of the country's terrestrial territory and 25 percent of its coastal waters, and conserves important land to sea connections (Centro Nacional de Áreas Protegidas 2013).

Cuba has since increasingly established protections and, as of 2020, 70 of the 105 marine protected areas (MPAs) have implemented some level of a management system (Valderrama et al. 2018). The protected area system safeguards around 30 percent of coral reefs, 24 percent of seagrass beds, and 35 percent of mangroves (Valderrama et al. 2018). Other national policies that oversee coastal development, protection, and ecosystem restoration also bolster marine conservation efforts (Whittle and Rey Santos 2006). Overall, Cuba's marine environmental policies are strong due to their foundations in science, operation across multiple ecological and spatial scales, and extensive coordination of management (Gerhartz et al. 2018).

But Cuba's ecosystems and biodiversity are facing serious threats, including the growing impacts of climate change; Cuba is already seeing rising sea levels and more frequent storms. Yet, adaptation, risk reduction, and recovery are possible. The country's extensive mangrove systems function as a protective barrier and scientists have found that some coral species are more resilient to disturbances and better able to recover from coral bleaching events (Caballero et al. 2019). In fact, Beyer et al. (2018) have identified the coral reefs around the world that have the best chance of surviving the impacts of climate change, with the view to help repopulate neighboring reefs. Of the fifty sites they identified, five are at least partially located in Cuba's waters, making the country home to one-tenth of the world's potentially most resilient reefs.

Cuba's important advancements in marine science, conservation, and policy will continue to benefit coastal communities through environmental services, such as storm surge protection, as well as through economic opportunities provided by healthy ecosystems including fishing and ecotourism.

WHY SHOULD THE UNITED STATES COLLABORATE WITH CUBA ON MARINE CONSERVATION?

Cuba and the United States share waters, marine ecosystems, and biodiversity. Ocean currents connect the two countries and wider regions of the Gulf of Mexico, Caribbean Sea, and Atlantic Ocean. Migratory species also connect the countries, including sharks, rays, tunas, billfish, sea turtles, and birds (García-Machado et al. 2018; Rooker et al. 2019). Cuba's marine ecosystems provide habitat to snappers and other reef fishes that spawn in Cuban waters, producing larvae that travel to nearby countries (Kough et al. 2016).

The two countries also face shared environmental threats, including climate change, pollution, oil and gas development, and overfishing. EDF collaborates with Cuba and countries around the world to collaboratively develop solutions to these challenges. Over the past decade, EDF's oceans program has prioritized recovering depleted fisheries by improving fisheries management to reduce the largest threat currently facing the world's oceans and coastal communities—overfishing. While Cuba's long history of marine research and conservation programs is highly advanced compared to other Caribbean countries, overfishing remains a serious problem for many species. Enacting adaptive fisheries reforms that make fishing sustainable can maintain or increase many fish populations in the future. This translates to more food and profits from fishing, even with the challenges presented by climate change (Gaines et al. 2018). Our shared environment and common threats make working together a necessity if we are to make marine conservation, including fisheries sustainability, successful.

SELECTED SUCCESSFUL INTERNATIONAL MARINE CONSERVATION COLLABORATIONS

International collaborations in marine conservation are already producing important results for nature and people in Cuba and around the world, and they are working because of diverse stakeholder participation. A selection of successes from collaborations that included EDF participation illustrate the benefits of working collaboratively, but these do not represent the full range of Cuba's extensive international environmental collaborations. Resource managers, policymakers, scientists, fishing industry representatives, conservation specialists, coastal communities, NGOs, and others continue to work collaboratively around shared goals for a sustainable and productive ocean. The positive outcomes described below were made by the strong commitment and active participation of a wide range of Cuban stakeholders.

- Cuba and the United States are now better prepared to prevent oil spills and to more effectively respond to future impacts. In 2017, after more than six years of scientific exchange as well as government and industry collaboration, the two countries signed a bilateral agreement to coordinate efforts in this arena (Agreement on Pollution 2017).
- Cuba's network of protected areas across land and coastal waters is one of the most ambitious and effective systems in the Caribbean. Part of the network was designed in collaboration with international organizations and experts. Cuba's success is instructive to scientists in the United States who are learning from Cuban partners' strategies to protect and restore coral reef ecosystems.
- In 2015, Cuba became the first Caribbean island nation to enact the National Plan of Action for the Conservation and Sustainable Management of Sharks & Rays (known as NPOA-Sharks), following the guidelines set forth by the United Nations Food and Agriculture Organization (MINAL 2015). The plan aims to develop the science and management actions needed to conserve healthy populations that support important tourism and fishing activities and marine ecosystems. Cuba's Ministry of Food Industry (Ministerio de Industria Alimentaria, or MINAL) led the plan, and EDF supported its development and implementation. Many species of sharks travel long distances, crossing borders (Hueter et al. 2017), and the development of the NPOA-Sharks brought together experts from, among others, Cuba, the United States, Mexico, Australia, Colombia, Ecuador, and Canada to share lessons learned and provide advice. The collaboration helped train fishers and port employees to collect new data on sharks that are caught around the country. This database allows scientists and managers to work with the fishing industry to protect vulnerable species.

- Through a four-year project called SOS Pesca, communities in Cuba became involved in science and natural resource management, found ways to fish more sustainably and developed new economic opportunities. The National Center for Protected Areas coordinated the project in partnership with the Italian NGO, COSPE Onlus, and the office of World Wildlife Fund–Netherlands (WWF–NL) in Cuba. SOS Pesca involved many other Cuban institutions and international organizations, including EDF, and achieved greater fisheries sustainability, conserved marine habitats, and improved the quality of life in two fishing communities in southern Cuba. The project achieved results at the local, regional fishing zone, and national scales. Through capacity-building workshops and international exchanges fishers, managers and scientists shared experiences with their counterparts from the United States, Mexico, and Colombia with the aim of making fishing more sustainable and developing community-based conservation programs (Miller et al. 2018).
- In 2018, Cuban and U.S. specialists published a scientific study that assessed the vulnerability of thirty-four commercially important fish species to overfishing in each of Cuba's four fishing zones (Puga et al. 2018). This study, initiated during the SOS Pesca project, grew out of a highly collaborative process that brought together scientific data and on-the-water expertise from the fishing industry to jointly assess fish populations (Miller et al. 2018). Managers are now using the results to prioritize the most vulnerable species for improved monitoring and management. During this broader collaboration and in the years following, Cuba implemented new regulations to improve management for lobster, sea cucumber, and the lane snapper fishery (CITMA 2019). Through limiting catch to more sustainable levels, the fisheries are stabilizing and, in some cases, improving.
- Partnerships have continued to grow since the conclusion of the official SOS Pesca project, and the project's success has spurred greater community participation in fisheries science and management. Fishers who participated in the SOS Pesca project have become ambassadors for sustainable fishing, marine conservation, and alternative livelihoods across the island, sharing their experiences with other communities. These activities have also created the opportunity for exchanges between Cuban scientists, managers, fishers, and dozens of their counterparts from the United States, Mexico, Belize, and Chile who share the same objectives. This partnership proposed a "Learning Network for Sustainable Fisheries in Cuba" that will continue to bring stakeholders together to share knowledge, tools, and lessons learned into the future (Morón et al. 2019).
- In 2018 and 2019, EDF partnered with Cuba's Center for Marine Research at the University of Havana and the Center for Fisheries Research to develop a new program to build the scientific and technical capacity of

professionals working in fisheries-related fields. Together the organizations designed and implemented two editions of a "University Course on Sustainable Fisheries Management" at the University of Havana that brought together working professionals (scientists, fishing port employees, fishing industry leaders, and conservationists) from all over the island. Participants drafted conservation-minded fisheries management plans for five fish stocks that were declining. Attendees improved their knowledge and skills (broadening tools to assess fish populations and to participate in developing future regulations) and built relationships with their peers that have already produced new partnerships and initiatives.

- In 2018, after fifty years of not publishing work by Cuban authors, the University of Miami's *Bulletin of Marine Science* published a collection of eighteen scientific articles in a Special Cuban Edition covering the newest research on marine ecology, fisheries, and conservation. EDF and partners at Vermont's Gund Institute persuaded the journal editors to reverse their decades-long policy in light of U.S. regulations that clearly authorized journals in the United States to publish articles from Cuban authors. Cuban and U.S. colleagues co-authored many of the articles, displaying the long and successful history of joint work on science and the environment (Whitefield 2018).

These selected results highlight the power of long-standing relationships built on science, shared values, and respect.

CUBA CONTINUES TO LEAD THE WAY ON CONSERVATION IN THE CARIBBEAN AND SET THE STAGE FOR GREATER COLLABORATION

In 2019, Cuba took proactive measures and announced new protections for the goliath grouper (*la guasa*), prohibiting all fishing of this species, which can grow up to 360 kilos. It is an important action for the whole region as the species is highly vulnerable and popular with scuba divers, particularly in Cuba's renowned Gardens of the Queen National Park. EDF has collaborated with scientists studying this species in Cuba, and their research has been an important resource for Florida where the goliath grouper is also protected.

In July 2019, Cuba's National Assembly passed the country's first Fisheries Law, replacing the previous Decree Law 164 of 1996 (Ley No.129/2019 Ley de Pesca 2020). The updated legislation ushers in a new era focused on ending overfishing, recovering depleted stocks, and strengthening the economies of coastal communities. The goal of the law is sustainable fisheries management in support of greater food sovereignty, security, conservation, and fishing jobs. It reflects years of work by Cuban institutions to improve science, management, and builds on lessons learned from international projects. One key provision of the law is its mandate that future management measures for all species must be based on science. This aligns with the U.S. federal fisheries law, setting the stage for continued opportunities for exchange and collaboration between the two countries toward achieving sustainable fisheries in shared waters.

THE IMPACTS OF CUBA'S SUCCESS IN MARINE CONSERVATION: REACHING OTHER COUNTRIES AND MOTIVATING ACTION THROUGH INTERNATIONAL COLLABORATIONS

EDF's work with partners on fisheries sustainability in Belize, Chile, Peru, the Philippines, Myanmar, China, and elsewhere is partly inspired by Cuba, which serves as a model of science-based conservation that involves communities. Recent examples include:

- In Chile and Peru, fishing stakeholders in each country developed their own fisheries learning networks that build on the lessons learned from the community-based SOS Pesca project in Cuba.
- In meetings in 2019, with partners in Myanmar, Cuba's assessment of its national fisheries was a leading global case study.
- In 2019, scientists from Cuba joined representatives from Belize, Mexico, the Dominican Republic, and the United States at the 72nd annual conference of the Gulf and Caribbean Fisheries Institute to share recent advances in effective fisheries management and policies across the region (Gulf and Caribbean Fisheries Institute 2019).

Cuban scientists have a long history of advising Belizean experts on fisheries science, and in recent years, EDF has created more opportunities for scientists, managers, and fishers in both countries to learn from each other (Miller 2018). In 2020, Belize also adopted a new national fisheries law, opening up further possibilities for Cuba and Belize to learn and work collaboratively as they implement new laws and continue to develop management plans and regulations.

International cooperation between Cuba and the United States on marine conservation, science, and the environment has illustrated the importance of ongoing and increased collaboration and the durability of this relationship. By broadening collaborations to involve countries throughout Latin America and the Caribbean, partners are able to share important lessons learned from Cuba. This resilient and broader collaboration is important for U.S.–Cuba relations and will be key to facing future environmental threats, such as the impacts of climate change, together.

Of course, many questions remain that continued international collaboration on marine conservation could help answer. These include:

- Why are Cuba's coral reefs more resilient to bleaching than many reefs in other places? What can other countries learn from Cuba's research to support their work to protect and recover coral reefs?
- What can data from MPAs tell us about the health of our ecosystems and our fisheries?
- What are the economic benefits of sustainable fisheries and conservation?
- What can Cuba do to recover fish populations, even in the face of climate change?

This last question is a particular EDF focus. The organization is working in over a dozen countries around the world and is committed to continuing collaboration with Cuba. EDF has also been working on a new project with MINAL, among other institutions, with the aim of supporting Cuba in creating new management plans for fish species that have declined and are critical to food security, employment, and healthy oceans.

FUTURE U.S.–CUBA ENVIRONMENTAL COLLABORATION: BASED ON COMMON GOALS

Collaboration in marine conservation is already achieving important results, as Cuba continues to make advancements in the field of sustainable fishing. Notably, the impact of these efforts is growing. Beyond these areas of collaboration, broader common goals provide an opportunity for expanding collaboration on science and the environment that can benefit both Cuba and the United States. Beyond marine conservation, environmental goals the two countries share include: (1) adopting clean energy solutions and fighting climate change and (2) adapting to climate change impacts by improving resilience in fisheries, farms, forests, energy systems, and communities.

Common Goal 1: Adopt clean energy solutions that cut greenhouse gas emissions to fight climate change, gain energy independence, and reduce air pollution for improved public health.

Cuba's commitment to fight climate change through mitigation and adaptation strategies is evident and was reaffirmed at the 2019 United Nations Climate Change Conference (COP25) in Madrid. Cuba is one of the few countries in the world whose constitution specifically mentions climate change. Energy independence is a policy priority, and the government has made ambitious commitments to adopt clean energy sources, move to a low-carbon economy, and it has already taken steps toward developing new solar and wind energy projects.

Although the United States arguably has a much greater responsibility to mitigate this crisis (as a much greater producer of global emissions), the Trump administration dismantled federal protections on clean air and water, environmental justice, and renounced U.S. participation in the 2015 Paris Accord on climate change mitigation. Nonetheless, while on the surface it may appear that the two countries shared no common ground during this period, it is not the case. Although there is no substitute for federal climate change and environmental legislation, city governments, state governments, NGOs, local citizen groups, students and private businesses have united around environmental protection in the United States. This has included arguing in courts, as well as taking individual actions to improve federal legislation, environmental protection, and climate change mitigation. Almost half of all U.S. state governors vowed to uphold the Paris Accord on climate change. For example, California and Hawaii have set deadlines for utilities to get 100 percent of their electricity from zero-carbon sources by 2045, and California is already ahead of the emission reductions target, while the economy continues to grow. Similarly, Puerto Rico intends to be 100 percent renewable by 2050. While action in California may be no surprise, Republican government-led states like Texas, which have economies built on oil production, are also becoming leaders in renewable energy. Indeed, EDF has worked with states across the country to reduce emissions, including in California, Texas, Colorado, Ohio, Illinois, New York, and Pennsylvania. These projects involve state and city policymakers, local organizations, public and private utilities, as well as many other partners outside of the federal government.

In addition to reducing emissions, air pollution is a concern for both the Cuban and the U.S. governments and populations. Air pollution is a global public health problem, disproportionately affecting marginalized communities, contributing to cardiac arrest, asthma, and other respiratory illnesses, and causing an estimated 4.2 million premature deaths annually around the world (WHO 2018). Partnerships at the city level to understand the problem of air pollution in Oakland, CA, Houston, TX, and London, UK, are deploying low-cost air quality monitoring sensors. With the help of Google and local organizations, sensors measure, and map residents' daily exposure to air pollution, using a street-by-street model. This data allows residents to see where the problem is most severe and helps cities make strategic decisions to reduce pollution and improve health where it is most needed (Environmental Defense Fund 2019). These types of initiatives that use technology to address public health and climate change priorities could benefit from Cuba's experience in health research and healthcare strategies to find new, creative solutions to reduce air pollution. Large cities in Cuba, like Havana, could benefit from collaborations with other urban centers and the new technology being used to tackle the same air pollution–based health issues.

While U.S. emissions are significantly greater than Cuba's, both nations are in many ways working toward the common goals of clean energy, clean air, and emissions reduction. The strategies to accomplish these goals are, and will likely continue to be, similar. These solutions include wind and solar energy, improving efficiency, electrification of transportation, energy storage and transmission, as well as public health initiatives. Cuba and the United States can benefit by working collaboratively through this process and developing partnerships with other countries. Although both Cuba and the United States have already made important advances, significantly more work is needed to reach their goals.

Common Goal 2: Adapt to climate change impacts by improving resilience in fisheries, farms, forests, energy systems, and communities.

Cuba has continued to excel at climate change adaptation. For decades, the country has been building the science and technology needed to plan for future scenarios and put strategies into place to reduce vulnerability to rising sea levels, increases in the intensity and frequency of storms, extended droughts and rising temperatures. Project Life (*Tarea Vida*), a national climate action plan being implemented across Cuba, continues to provide important examples and lessons for the United States as well as the global community. For example, one particular area that could benefit from a greater collaboration is Puerto Rico, which has continued working to rebuild and improve its energy systems to withstand and recover from future storms. When Hurricane Maria hit the island in 2017, it caused major loss of life. In addition, it destroyed 80 percent of the island's power lines, leaving some residents without electricity for over a year. EDF is working with local communities and experts in Puerto Rico to develop resilient solar energy systems that can deliver clean, affordable power to rural areas, even during a storm (Carbó 2020). This work could significantly benefit from Cuba's experience in disaster preparedness, relief, adaptation planning, and distributed energy systems. Collaboration aligns with Cuba's commitment to clean energy and could spur new solutions for the Caribbean.

Notably, knowledge exchange has been occurring in other fields for many years, particularly around agroecology. Cuba's National Association of Small Farmers (Asociación Nacional de Agricultores Pequeños, or ANAP) holds biannual international meetings, which attract participation from U.S. farmers to learn and exchange best practices. A newer collaboration addresses the interconnectedness of environmental sustainability across economic sectors; the Research

Initiative for the Sustainable Development of Cuba (RISDoC) is a coalition of Cuban and international NGOs, academics, government agencies, and others working together to advance and share knowledge on a range of issues relating to sustainability, climate resilience, and Cuba's participation in the global low-carbon economy. In February 2017, RISDoC hosted its first seminar in Havana, which included presentations and discussions around environmental sustainability in key sectors, such as agriculture, energy, and tourism. The Foundation of Antonio Núñez Jiménez, Caribbean Agroecology Institute, and EDF serve as co-chairs of this initiative and help connect experts in Cuba and across the Caribbean. RISDoC continues to facilitate dialogue and exchange around pressing environmental issues that are pertinent to Cuba, the United States, and the wider Caribbean region.

CONCLUSION

EDF will continue to work with Cuban institutions on sustainable fisheries science and management and support broader environmental collaboration around land-to-sea conservation issues. In addition to commercial fishing, well-managed and equitable recreational fishing tourism presents an opportunity for economic growth that could build on the deep knowledge and experience of fishers around the country and Cuba's rich marine biodiversity.

The list of potential areas of collaboration on science and the environment is long. In the short term, importance continues to be placed on developing collaborations and building relationships, including engaging in opportunities in third-party countries. Ongoing collaboration is crucial; when U.S. and Cuban diplomatic relations improve (again), there will already be a solid foundation of shared environmental goals on which to build. This is, and will continue to be, necessary to create stronger and more successful partnerships for achieving environmental sustainability and marine conservation.

BIBLIOGRAPHY

Agreement on Pollution. 2017. US-Cuba Cooperation agreement on preparedness for and response to pollution caused by spills of hydrocarbons and other noxious and potentially hazardous substances in the Gulf of Mexico and the Straits of Florida. Signed in Havana on January 9, 2017. Ratified June 30, 2017.

Caballero Aragón, Hansel, Maickel Armenteros, Susana Perera Valderrama, Néstor Rey Villiers, Dorka Cobián Rojas, Karla Campos Verdecia, and Pedro M. Alcolado Menéndez. 2019. "Ecological Condition of Coral Reef Assemblages in the Cuban Archipelago." *Marine Biology Research* 15 (1): 61–73.

Carbó, Agustín, and Amalia Saladrigas. "Resilience in the Eye of the Storm: How Puerto Rico Can Build a Stronger, More Sustainable Energy Future." *Environmental Defense Fund,* June 30. Accessed April 28, 2020. http://blogs.edf.org/energyexchange/2020/06/30/resilience-in-the-eye-of-the-storm-how-puerto-rico-can-build-a-stronger-more-sustainable-energy future/?utm_source=twitter&utm_campaign=energyex_none_upd_engy&utm_medium=social-media&utm_id=1593533287.

Centro Nacional de Áreas Protegidas. 2013. *Plan del Sistema Nacional de Áreas Protegidas 2014–2020*. Havana.

CITMA. 2019. *Sexto Informe Nacional al Convenio Sobre la Diversidad Biológica República de Cuba.* Ministerio de Ciencia, Tecnología y Medio Ambiente. May. Accessed April 28, 2020. http://www.inaf.co.cu/sites/default/files/inline-files/6IN%20FINAL-1.pdf.

Environmental Defense Fund. 2019. *Making the Invisible Visible: A Guide for Mapping Hyperlocal Air Pollution to Drive Clean Air Action.* Accessed April 28, 2020. https://www.edf.org/sites/default/files/content/making-the-invisible-visible.pdf.

Gaines, Steven D., Christopher Costello, Brandon Owashi, Tracey Mangin, Jennifer Bone, Jorge García Molinos, Merrick Burden, Heather Dennis, Benjamin S. Halpren, Carrie V. Kappel, Kristin M. Kleisner and Daniel Ovando. 2018. "Improved Fisheries Management Could Offset Many Negative Effects of Climate Change." *Science Advances* 4 (8): eaao1378.

García-Machado, Erik, Gabriela Ulmo-Díaz, Jessy Castellanos-Gell, and Didier Casane. 2018. "Patterns of Population Connectivity in Marine Organisms of Cuba." *Bulletin of Marine Science* 94 (2): 193–211.

Gerhartz-Muro, José L., Jacob P. Kritzer, Adrián Gerhartz-Abraham, Valerie Miller, Fabián Pina-Amargós and Daniel Whittle. 2018. "An Evaluation of the Framework for National Marine Environmental Policies in Cuba." *Bulletin of Marine Science* 94 (2): 443–459.

González-Díaz, Patricia, Gaspar González-Sansón, Consuelo Aguilar Betancourt, Sergio Álvarez Fernández, Orlando Perera Pérez, Leslie Hernández Fernández, Víctor Manuel Ferrer Rodríguez, Yenisey Cabrales Caballero, Maickel Armentros Almanza, and Elena de la Guardia Llanso. 2018. "Status of Cuban Coral Reefs." *Bulletin of Marine Science* 94 (2): 229–247.

Gulf and Caribbean Fisheries Institute. 2019. *Fishery Leaders from Gulf of Mexico, Caribbean Meet to Tackle Overfishing and Climate Impacts.* November 6. Accessed April 28, 2020. https://www.gcfi.org/download/59/gcfi72/6370/gcfi-edf-press-release.pdf.

Hueter, Robert E., John P. Tyminski, John J. Morris, Alexei Ruiz Abierno, and Jorge Angulo Valdes. 2017. "Horizontal and Vertical Movements of longfin makos (Isurus paucus) Tracked with Satellite-linked Tags in the Northwestern Atlantic Ocean." *Fishery Bulletin* 115 (1): 101–116.

Kough, Andrew S., Rodolfo Claro, Kenyon C. Lindeman, and Claire B. Paris. 2016. "Decadal Analysis of Larval Connectivity from Cuban Snapper (Lutjanidae) Spawning Aggregations based on Biophysical Modeling." *Marine Ecology Progress Series* 550: 175–190.

Kritzer, Jacob P., Crtistina C. Hicks, Bruce D. Mapstone, Fabian Pina-Amargós, and Peter F. Sale. 2014. "Ecosystem-based Management of Coral Reefs and Interconnected Nearshore Tropical Habitats." In *The Sea, Volume 16: Marine Ecosystem-Based Management*, edited by Michael J. Fogarty and James J. McCarthy. Cambridge: Harvard University Press.

Ley No. 129/2019 Ley de Pesca. 2020. *Gaceta Oficial de la República de Cuba*. February. Accessed April 15, 2020. https://www.gacetaoficial.gob.cu/es/decreto-1-de-2019-de-consejo-de-ministros.

Mancina, Carlos A., and Daryl D. Cruz Flores, editors. 2017. *Diversidad biológica de Cuba: Métodos de inventario, monitoreo y colecciones biológicas*. Havana: Editorial AMA.

Miller, Valerie. 2018. "What do Cuban and Belizean Fishers Have in Common? More than Meets the Eye." Environmental Defense Fund. August 16. Accessed April 28, 2020. http://blogs.edf.org/edfish/2018/08/16/what-do-cuban-and-belizean-fishers-have-in-common-more-than-meets-the-eye/.

Miller, Valerie, Ania Mirabal-Patterson, Elisa García-Rodríguez, Kendra Karr, and Daniel Whittle. 2018. "The SOS Pesca Project: A Multinational and Intersectoral Collaboration for Sustainable Fisheries, Marine Conservation and Improved Quality of Life in Coastal Communities." *MEDICC Review* 20 (2): 65–70.

MINAL. 2015. *Plan de Acción Nacional de Conservación y Manejo de Condrictios de la República de Cuba*. Ministerio de la Industria Alimentaria. Havana: Ministerio de la Industria Alimentaria.

Morón, Eduardo Boné, Raidel Borroto Vejarano, and Valerie Miller. 2019. "Cuba: Netting Fishers." *Samudra Report* 81: 4–7.

Puga, Rafael, Servando Valle, Jacob P. Kritzer, Glima Delgado, María Estela de León, Enrique Giménez, Ibis Ramos, Orestes Moreno, and Kendra Karr. 2018. "Vulnerability of Nearshore Tropical Finfish in Cuba: Implications for Scientific and Management Planning." *Bulletin of Marine Science* 94 (2): 377–392.

Rooker, Jay R., Michael A. Dance, R. J. David Wells, Matthre J. Ajemian, Barbara A. Block, Michael R. Castleton, J. Marcus Drymon, Brett J. Falterman, James S. Franks, Neil Hammerschlag, Jill M. Hendon, Eric R. Hoffmayer, Richard T. Kraus, Jennifer A. McKinney, David H. Secor, Gregory W. Stunz, and John F. Walter. 2019. "Population Connectivity of Pelagic Megafauna in the Cuba-Mexico-United States Triangle." *Scientific Reports* 9 (1): 1–13.

Valderrama, Susana Parera. Aylem Hernández Ávila, Juliett González Méndez, Orestes Moreno Martínez, Dorka Cobián Rojas, Hakna Ferro Azcona, Elvis Milián Hernández, Hansel Caballero Aragón, Pedro M. Alcolado, Fabián Pina-Amargós, Zaimiuri Hernández González, Leonardo Espinoa Pantoja, and Lázaro Francisco Rodríguez Farrat. 2018. "Marine Protected Areas in Cuba." *Bulletin of Marine Science* 94 (2): 423–442.

Whitefield, Mimi. 2018. "U.S., Cuban Marine Biologists Put an End to 'academic embargo.'" *Miami Herald*. October 20. Accessed April 28, 2020. https://www.miamiherald.com/news/nation-world/world/americas/cuba/article219965430.html#:~:text=There%20are%20no%20borders%20that,to%20get%20along%20and%20collaborate.

Whittle, Daniel J., Kenton C. Lindeman, and James T.B. Tripp. 2002. "International Tourism and Protection of Cuba's Coastal and Marine Environments." *Tulane Environmental Law Journal* 16: 533–589.

Whittle, Daniel, and Orlando Rey Santos. 2006. "Protecting Cuba's Environment: Efforts to Design and Implement Effective Environmental Laws and Policies in Cuba." *Cuban Studies* 37: 73–103.

WHO. 2018. *Ambient Air Pollution: Health Impacts*. World Health Organization. September 15. Accessed April 18, 2020. https://www.who.int/airpollution/ambient/health-impacts/en/.

Chapter 11

Agroecology, Food, and the Climate Crisis

Transition, Adaptation, and Building Resilience in Cuba

Margarita Fernández, Leidy Casimiro Rodríguez, Luis L. Vázquez, and Giraldo Martín Martín

The Caribbean and Cuba especially are ground zero for the impacts of the global climate crisis where temperatures are increasing, sea levels are rising, coral reefs are bleaching, droughts are becoming longer and more severe, and hurricanes are becoming more intense and frequent. As the climate crisis intensifies, the urgency to respond is becoming more critical. One of the greatest challenges we face as a global community is how to improve the quality of life, while adapting to, and mitigating, the climate crisis, as well as preserving the finite natural resources on which society and the economy depend. The transformation of the agrifood systems is essential to achieving this challenge. Evidence shows that the industrial agrifood model has failed to adequately and nutritiously feed the population while perpetuating a system that is ecologically destructive and socioeconomically unjust, yet politically powerful. It contributes up to 37 percent of greenhouse gases and is poisoning and damaging our natural resources base as well as us as human beings (IPCC 2019). The COVID-19 global pandemic has unveiled the vulnerabilities of a monocultural industrialized global food system, highlighting the interdependence of human, animal, and environmental health. It has also illustrated the central role of what and how we produce, distribute, and consume agrifood products.

Key to building resilience at different levels is decreasing our use and dependency on fossil fuels, increasing the amount of carbon sequestered, and increasing our absorptive, adaptive, and transformative capacity in social, economic, and ecological dimensions to better respond to shocks and stressors. In Cuba, the lived experience of resilience runs deep, as a society that has withstood the many stresses and shocks of colonialism, the U.S. blockade, the Special Period, and more recently the effects of climate change. Cuba's advances in agroecology go hand-in-hand with their ethos of resiliency. The combination of strong social networks and government support for disaster preparedness and response, with agroecological farms in Cuba has resulted in greater resilience for those farms and communities in the face of increasing effects of droughts, floods, and hurricanes, as well as socioeconomic shocks and stressors, as documented by several Cuban researchers (Casimiro Rodríguez 2016; Machín Sosa et al. 2010; Vázquez Moreno et al. 2016).

Agroecology is integral to building climate resilience as an agrifood systems paradigm that more effectively addresses the problems of hunger, poverty, and inequality while curbing biodiversity loss and providing more resilience to climate change. In other words, agroecology is the application of ecological principles to the design and management of diversity in agroecosystems through the integration of traditional and scientific knowledge (Altieri 1995). However, agroecology is not just a science and on-farm practice but also a social and political movement that seeks to transform the dominant corporate food model by researching, implementing, and advocating for socially just, economically fair, and ecologically resilient models (Giraldo and Rosset 2017). Cuba actualizes many of the central principles of agroecology, from low-external-input production to diversified farming systems and horizontal learning methodologies. Across the hemisphere, the Cuban case stands large as an extraordinary instance of revolutionary agrarian reform against improbable odds. Cuba has promoted and implemented agroecology transitions for over twenty years as part of its struggle to sustain national food security and domestic sovereignty.

Although Cuba faces many challenges in its agriculture and food system, since the early 1990s, sparked by the fall of the Socialist Bloc, they have been developing key building blocks for a resilient agrifood system through the decentralization of large state farms to smaller units of cooperative production, the redistribution of land to small farmers, the decentralization of food systems planning and distribution to the municipal level, a laudable national urban agriculture program, the promotion of diversity on farms, the integration of alternative renewable energy sources on farm, and a farmer-to-farmer extension program rooted in agroecology.

While the movement was born out of necessity, there was a cadre of knowledgeable farmers and scientists who were practicing, researching, and advocating for more sustainable practices for years. These individuals were able to take advantage of the opening of the political space to lead in the practice, science and policy of agroecology. Over time, more farmers, scientists, and decision-makers began to understand the destructive practices of their previous industrial model and the high level of foreign dependence, vulnerability, and risk that the monoculture, export economy placed on their nation. While there are still many in Cuba who believe they can only solve the country's food security challenges through high-tech, high-capital, large-scale production systems, in recent years, the pendulum has been swinging again—this time in the direction of agroecology. This has very much been in response to the climate crisis and the need to build a robust and resilient food system. In particular, there has been renewed urgency to make profound reforms as a result of the intensifying economic and food crisis, sparked by the tightening of the blockade under the Trump administration, the crisis in Venezuela, and the impacts of COVID-19 on tourism.

In April 2020, Cuba announced a national strategy for food sovereignty and nutritional education, which is aligned with the national climate change strategy *Tarea Vida*. This new plan is an important milestone for the agroecology and food sovereignty movements and will help pave the way for actors in Cuba and the international cooperation sector, to bolster the advances Cuba has already made in the past several decades. Many of the challenges to the island's agriculture and food system have less to do with whether the mode of production is agroecological or conventional, though this plays a role or levels of productivity; rather, challenges are based largely within structural, legal, and economic constraints. This includes the severe limitations imposed by the U.S. blockade, scarcity of inputs (i.e., hand tools, solar panels, water pumps, fencing, processing equipment, etc.), lack of diverse markets to sell products, a high rate of food waste due to inefficient distribution systems and lack of adequate processing and storage facilities, among others (Fernandez et al. 2018).

This chapter presents empirical data from Cuba to highlight three important elements of resilience in the island nation's food system. First, results from a socioecological analysis of family farms in the cooperative sector that outline the successes, challenges, potentials, and policy recommendations are presented. Second, the results from an ongoing national project that integrates energy production with food production on farms in order to stem reliance on fossil fuels, decreasing carbon emissions, and increasing the energy sovereignty and resilience of Cuba are offered. Lastly, results are shared from an experience with agroecological co-innovation for drought resilience functions. Droughts in Cuba are one of the most detrimental impacts of climate change, and as such, strategies to adapt and build resilience are paramount to ensuring food security and sovereignty on the island.

THE ROLE OF FAMILY FARMING IN STRENGTHENING RESILIENCE

Family farms are recognized mainly as the place of residence of peasant families and providers of the vast majority of the labor force in food production, notwithstanding a family's occasional employment in other activities or its hiring of temporary labor. Despite covering only 20 percent of agricultural land, family farming feeds more than 70 percent of the world's population (FAO 2017). This is why it is important to understand rural territories by listening, observing, recognizing tastes, traditions, and needs of communities, not only to highlight their role in food production and ecosystem services but also to strengthen their socioecological resilience. Resilience involves the capacity to carry out adaptive changes in the face of disturbances or situations of stress and to remain robust over time, maintaining agricultural production in harmony with rural families' culture, social organization, needs satisfaction, and the ecosystem's capacity, maintaining respect and ethics in the stewardship of the land, while assuring peasant wellbeing.

In Cuba, peasant families own approximately 9.1 percent of the country's total surface area; this translates to 875,000 hectares of agricultural land (or 13.9 percent of the total agricultural land). However, as of 2008, these numbers have increased as a result of Laws 259 and 300, which distribute unused arable land to farmers under usufruct rights. Most of these farms are grouped into credit and service cooperatives,[1] which produce more than 70 percent of Cuba's food (ONEI 2018, 2019). As a result of the different modalities of usufruct, urban, and suburban agriculture, new families have been incorporated into the food production process, bringing with them a new understanding of family agriculture.

The importance of family farming in Cuba is based not only on the diversity and quantity of food produced historically but also involves experimentation and peasant innovation to confront the different technological, financial, and market limitations. This has resulted in a multidisciplinary peasant population capable of assuming different roles and high levels of efficient productivity in comparison with other models of agricultural production in the country. Studies conducted in more than thirty rural territories covering provinces in the three main regions (East, Central, and West) in Cuba, using a methodology to assess socioecological resilience of family farms, illuminate the efficacy of this agricultural model (Casimiro Rodríguez 2016). Specifically, they highlight characteristics and common features of Cuban peasant agriculture that make it more resilient and less vulnerable to climate change and market fluctuations. For example, one characteristic identified is the presence of mixed agricultural and livestock production systems, where producers manage and conserve an important diversity of seeds, breeds, and cultivated varieties. This includes several agroecological practices for soil conservation and improvement, polyculture, and silvopastoral systems, and less dependence on oil and its derivatives. Most of these farms have innovative working tools, designed and implemented by families themselves, some of which are used for animal traction in agricultural work and for local transportation. The energy efficiency of these systems is greater than 3 MJ of energy produced per MJ of energy consumed. Farms with a more advanced agroecological transition can produce up to twelve times more energy than it consumes (Casimiro Rodríguez 2016; Casimiro Rodríguez and Casimiro 2018; Funes Monzote 2018).

In general, family members largely supply the necessary farm labor, while also generating some employment in their communities. These families produce food for self-consumption and for the market, in a diversified manner, conserving local and regional biodiversity, making little use of agrochemicals. Farms that are in a more advanced stage of agroecological transition do not use synthetic chemicals in production, thus directly producing healthier foods for the population, while increasing the availability and consumption of fresh food in the region. The average production of these farms is 7 tons per hectare per year, and they are capable of feeding eight people per hectare per year in protein and energy (Casimiro Rodríguez 2016; Casimiro Rodríguez and Casimiro 2018).

Throughout the countryside, there are farmers leading in agroecological innovations that are linked to research centers and universities as well as participating in multiactor decision-making spaces through municipal platforms. Several of them are supported by international cooperation projects with training and infrastructure improvement activities for agroecological transitions. These projects have supported the implementation of different technologies for the use of renewable energy sources and the manufacture of organic fertilizers. Specifically, these projects include, among others, the construction and management of biodigesters, windmills, worm bins, implementation of composting systems, and bio-carbon. In addition, they preserve some foods, mainly tomato pulp, and some fruits, using natural and artisanal processing methods that help reduce post-harvest losses.

In spite of these achievements, there are several factors that put family farming at risk. Rural to urban migration continues, and with a rural population that is rapidly aging, there is a danger that younger generations will not replace the older ones, as they are often reluctant to continue living in the countryside. The ongoing effects of climate change are increasingly evident too, with extensive periods of drought. There has also been a loss of rural cultural traditions, such as *guateques* (community parties), fairs, rodeos, as well as the disappearance of rural basic trades, such as blacksmiths, carpenters, saddlers, and palm cutters.

In general, the families studied did not add value to what they produced, beyond the fruit pulp produced by some. This was mostly due to the scarce material resources for the processing and slaughtering of their products, in addition to not having the legal backing to directly market new products. This continues to affect peasant families' local economies, since they sell their unprocessed production at low prices. Were they able to add value to their products, they could lengthen their production cycles, generate self-employment for the family, and improve their income. In addition, this would contribute to the domestic market, which is still undersupplied and help decrease food waste.

Other factors that affect the resilience of farms include, among others, inadequate livelihoods (reflected in poor housing infrastructure), insufficient neighborhood roads and highways, little access to public transportation or to the market for agroecological inputs, the significant distance from schools and hospitals, and economic analyses that do not reflect family wages. Thus, practical actions to strengthen peasant agriculture are urgently needed. For example, programs, projects, and public policies that support the resilience and livelihoods of family agriculture based on agroecology should be focused on increasing efficiency in the use of endogenous resources, to encourage the incorporation of new farming families, to contextualize appropriate technologies for the maximum use of renewable energy sources, and to contribute gradually to the achievement of food sovereignty.

Local and national agricultural production must be protected and regulated so that domestic markets and distribution systems respond to the goals of self-sufficiency of food that is safe, nutritious, and promotes the multifunctionality of peasant production methods. Resources that are currently used to subsidize the purchase of technological packages for conventional agriculture, food distributed in the ration system, and the electrification of some homes could be reallocated to promote the growth of family farming based on agroecology. Many of the farms studied are more than 75 percent self-sufficient in food and have an energy consumption in their homes that is less than 60 percent of the average Cuban home (Casimiro Rodríguez and Casimiro 2018). With the use of biogas generated in their biodigesters, many of these farms obtain the energy necessary for cooking food, refrigeration, and lighting, while closing the nutrient cycle by using waste for organic fertilization of their crops. These resources could be used for the implementation of agroenergy[2] farms, such as those where technologies and innovations are implemented, improved, and systematized to integrally produce food and energy, mainly from renewable sources. In turn, this would contribute to efficiency and socioecological resilience, while using locally available resources.

Funds could also be used to support short supply chains that retain value in the community, reduce physical distances and transaction costs, and educate the wider public about solidarity and responsible consumption. Food systems rooted in a given region highlight the understanding that food production and consumption is a farm-to-table process which contributes to the preservation of traditional agrifoods (agriculture plus kitchen-based knowledge), the reduction of food waste, an increase in food diversity, as well as improved knowledge of food sourcing and nutrition. Achieving integrated actions and public policies for understanding and promoting the links between peasant culture and socioecological resilience requires inclusive multisectoral strategies. This multisectoral approach would make it possible to raise awareness and encourage the population as a whole to preserve and promote family farming, improve production and nutrition, while recognizing the importance of these factors in achieving food sovereignty.

INTEGRATING FOOD AND ENERGY PRODUCTION FOR RESILIENCE AND SOVEREIGNTY

In rural areas, food production integrated with energy production from renewable sources is key to achieving sustainable and resilient agriculture capable of resisting the different problems that climate change inflicts on the farming sector, both at home and abroad. Energy from biomass produced at the farm level can meet local demands for energy in a way that is renewable and contributes to the mitigation of, and adaptation to, climate change. Since 2009, the Indio Hatuey Experimental Station of Pasture and Forage (known by its Spanish acronym, EEPF-IH), at the University of Matanzas, has coordinated a project called *BIOMAS: Biomass as a Renewable Source of Energy in the Rural Environment*, with funding from the Swiss Agency for Development and Cooperation (SDC) as well as the support and participation of various national and local institutions.[3] This project is being implemented in three successional phases that build on each other. Phase 1 (2009–2012) focused on developing pilot experiences in three thematic areas: biodiesel, biogas, and biomass gasification. Phase 2 (2013–2017) focused on developing a municipal strategy for integrated food and energy production. Phase 3 (2018–2021) focuses on improving residents' living standards in the municipalities participating in the program based on the consolidation and experiences of the previous phases. By 2020, some key results of the project include:

- Implementation of technological bioenergy solutions in 240 demonstration scenarios in eight municipalities that have directly benefited 21,983 people, created 480 new jobs (26 percent of which are held by women), improved and/or reforested 4,332 hectares of land, increased food production by 45 percent, generated $88.7 million in these scenarios, and produced 12,867 tons of biofertilizers.

- Strengthening of institutional, technical, and productive infrastructure in 265 national and local entities and 6,702 key actors (including producers, specialists, and decision-makers)—43 percent of which are women—have improved their technical and management skills.
- Preparation and implementation of local strategies for food and bioenergy production in six municipalities, in cooperation with the government and other local actors, as well as a notable impact on national and sectoral public policies (PPI), including the implementation of a national biodiesel program.
- Creation of skills for the production and use of biogas, biodiesel, and biomass gasification in eighteen municipalities (245 biodigesters, 3 biodiesel plants, 3 gasifiers, 52 mini bio-product and bio-fertilizer plants); 4,123 people benefit from biogas for domestic use, saving 388,800 kW.h/year and reducing household electricity consumption by 40–70 percent.
- Promotion of three international projects, with funding from the European Union (EU), Global Environment Facility (GEF), United Nations Development Program (UNDP), and Sweden, as well as three national projects.
- Creation of networks of national, sectoral, and local actors that have established synergies and alliances.
- Recognition by senior officials of the national and local governments, ministries, and producers.
- Coordination with other projects supported by the SDC, GEF, and UNDP.

A key achievement of the project to date is the establishment of micro-networks[4] that supply biogas to homes in rural communities of the municipality of Cabaiguán, in the province of Sancti Spíritus. These micro-networks have been established in four communities, benefiting 102 households that are saving 146 MW.h/year. More micro-networks are being built to benefit 82 households for a savings of 134 MW.h/year.

In regards to the production of biogas from livestock manure, an industrial biodigestor, with a volume of 4200m,^{3} was built in the Martí municipality of the province of Matanzas. This biodigestor has the potential to produce 3800–4000m^3 of biogas/day but is only producing 1500m^3/day as a result of the number of animals on the farm. With an electric generator attached, this biodigestor could generate up to 2.5 MW/day. As a byproduct, the system produces 24 tons of solid fertilizer per month that can be used on farms. When this biodigestor is at full capacity, part of the biogas could be used to produce biomethane to fuel the public transport system in the municipality (Martin 2019).

In addition, three biomass gasifiers have been installed. One of these uses sawmill wood waste, while another uses prunings from the silvopastoral and forest systems to generate electricity. The third is in a rice-processing plant that uses rice husks to produce a synthetic gas that is employed to dry the cereal grain. Any surplus gas may be used to generate electricity. This gasifier saves 91 tons of diesel consumed annually for drying rice, as well as 157 MW.h/year of electricity consumed at the plant, in addition to generating 442 MW.h/year for the national electricity system, by processing 1,186 tons of husk per year.

The use of these technologies leads to a rational and efficient use of locally available resources, furthering the goal of establishing closed energy and nutrient cycles that contribute to sustainable agricultural development at the municipal level. Both biodigesters and biomass gasifiers are alternatives that make it possible to convert waste into new products (such as gaseous fuels for different farm tasks or services, including cooking, refrigeration, electricity, and power generation). These systems in turn generate other wastes that can be used as a source of nutrients to fertilize crops and replenish degraded soils (Suárez 2019).

The production of liquid biofuels is another technology that can be implemented using mainly inedible oilseed plants (Sotolongo et al. 2019). By intercropping food crops with *Jatropha curcas* (an oilseed plant), food and energy security and sovereignty can be achieved. Research from trials on farms in the province of Matanzas demonstrates the capacity to produce 1.33 tons/hectare of food per year along with 1.6 tons of oilseed fruit per hectare per year in the same space.

Oilseed plants, such as *Jatropha curcas*, interspersed with crops or planted in rows and as live fences (1.2 kilometers of live fence with *Jatropha* is equivalent to one hectare with 833 plants within the farms), can help strengthen an energy security scheme. This translates to allocating no more than 30 percent of the land to produce the necessary liquid fuels and tending to crops on 100 percent of the available land. This model is best suited for small-scale rural industries, managed as a cooperative or state enterprise, to produce all or part of the fuel required to develop all crop-related tasks, such as soil preparation, planting, cultivation, harvesting, processing, and marketing.

So far, the results achieved with *Jatropha curcas* have not met the expectations, making it impossible to fulfill the goals proposed for biodiesel production (345 kg/ha/year or 394 liters/ha/year). The yields of harvested fruits and seeds are still far from what the international literature says is optimal (more than 4000 kg of fruits/ha/year), since it has not been possible to exceed 1691 kg of fruits/ha/year under current production conditions.

In the face of lofty promises, multiple disappointments, abandonment of crops, and bankruptcy (FAO-HLPE 2013), those involved in *Jatropha* initiatives have come to better understand the necessary caution when reporting planting results in order to produce realistic feasibility analyses. International cooperation agencies, such as the Netherlands Agency, GTZ (now GIZ) of Germany, and the United Kingdom's Department for International Development (DFID), commissioned critical analyses of *Jatropha* projects (GTZ 2009; Mubonderi 2012; van Eijck et al. 2010). The reported yields are low, representing approximately less than a tenth of what is needed to reach the expected biodiesel yields per hectare.

It is therefore necessary to rethink *Jatropha*-based projects, because the studies conducted in other countries were seemingly very optimistic. This species lacks genetic studies and shows great variability in flowering and fruit formation in the field. Currently, in 2021, research is being carried out to study the most suitable soils, the biology of flowering, fertilization and irrigation needs, and appropriate pruning, among other important aspects. More than seventy research, development and innovation activities related to the agro-industrialization of *Jatropha* have been identified in coordination with national universities in Cuba (Relatoría 2019).

In terms of a strategy to find an alternative that will advance local energy security of agricultural systems, it is important to consider studying other species of oilseed plants to evaluate their potential oil and biodiesel production. This would thus generate a diversified model that incorporates possible energy-producing crops, trees, and palms (such as *Corojo*, *Higuereta*, *Moringa*, *Aleuritis*, and *Neem*). In the meantime, other technologies must be incorporated, such as solar pumps, wind pumps, small wind turbines, and pico hydroturbines. Despite having different purposes, the primary aim remains centered on achieving energy sovereignty and contributing to the sustainable development of Cuba's rural environment.

AGROECOLOGICAL CO-INNOVATION FOR RESILIENCE TO DROUGHTS

Droughts are extreme weather events that have a significant impact on agricultural production. However, drought management strategies, even technological ones, rarely go beyond recommending that the water supply be guaranteed, rationed, and sustainably sourced and that agro-climatic zoning be considered in planning. In Cuba, agricultural production systems that are undergoing an agroecological transition are showing a greater capacity of resilience to droughts. At the same time, conventional systems prone to droughts are adopting agroecological practices to reduce their vulnerability. Evidence from these farms has helped bolster agroecology as a scientific and methodological approach to shift vulnerable systems to resilient ones, particularly because agroecological design and management contribute to resilience functions (Vázquez et al. 2019). In this section, results from three participatory evaluations conducted across the island are presented.

In many areas of the world, peasants have developed agricultural systems based on local conditions. This has allowed them to continually produce for subsistence, despite growing their crops on marginal land, with unpredictable climate variability, and very low use of external inputs (Altieri and Nicholls 2009). In Cuba, peasant agriculture is an example of an agroecological transition on a national scale made possible in large part to the *Movimiento Agroecológico de Campesino a Campesino* (Peasant-to-Peasant Agroecological Movement) organized by the National Association of Small Farmers (Asociación Nacional de Agricultores Pequeños, or ANAP). Farms that participate in an agroecological transition adopt a diversity of practices that are rooted in agroecological principles and have demonstrated greater capacity to resist, recover, and transform in the face of climate change events, such as hurricanes and droughts (Machín Sosa et al. 2010).

In order to better understand the vulnerability of production systems to climate change and to systematize innovative adaptive practices, a participatory analysis with technicians and farmers was carried out in ten agricultural territories in Cuba (Vázquez et al. 2015). A workshop was held in every territory; 95 people participated in the western region, 85 in the central region, and 86 in the eastern region, for a total of 266 participants. Participants were specialists from municipal agencies (including soils, plant health, veterinary services, agricultural production, and extension workers), technicians and managers of production organizations (cooperatives and farms), and agricultural producers (administrators and owners of farms and other production systems). Teams were organized at every workshop and received a printed guide with two sections: (a) Vulnerability: types of production systems, sensitivity criteria, and an assessment scale and (b) Adaptive Practices: purposes of the practices, which they used to evaluate the following types of production systems, predominant in the country: conventional agricultural and livestock farms, integral peasant farms, agroforestry coffee systems, *organoponicos* (raised bed production systems), and intensive urban agricultural gardens. The following components of an agroecosystem that are most exposed to the effects of drought were assessed: soil, crops in growing phase, crops in harvesting phase, water, and pest control. In order to identify the adaptive practices that are carried out by farmers in the territory, these were grouped according to their main resilience functions: (1) Capture, conserve, and optimize water; (2) Maintain soil moisture; (3) Reduce impact on soil properties; (4) Reduce effects of excess direct solar radiation on crops; (5) Reduce effects of pests and diseases; (6) Efficiently apply pesticides (chemical, biochemical, and biological) and release entomophages; (7) Reduce effects on pre-harvest production; and (8) Reduce post-harvest effects. Each team identified the practices implemented in their territory and classified them according to their purpose.

As a result of their diagnosis, participants concluded that more than 64 percent of the territories have a high level of vulnerability to drought, and conventional livestock systems have the highest level of vulnerability (90.4 percent of the territories). The diagnosis also identified 154 drought adaptation practices, most of which were focused on reducing the effects of pests and maintaining soil humidity by increasing the spatial, species, and genetic diversity on the farm (Vázquez et al. 2015). This process highlighted the diversity of practices and resilient functions being practiced due to the innovation of technicians and farmers and contributes to the improved design and management of agroecosystems for drought resilience using agroecological principles.

In 2016, a series of three workshops were carried out in the province of Guantanamo in order to create strong methodological foundations for the management of resilience. This province was prioritized, because it is the region in the country most exposed to drought (Planos et al. 2013; Solanos et al. 2000). Each workshop lasted five days and included presentations, group activities, on-farm evaluations, surveys with local actors, as well as other activities (Vázquez Moreno et al. 2016). Thirty-one people participated directly in the workshops (government officials, specialists, researchers, teachers, technicians, and farmers) from ten provincial and municipal level institutions. The breadth of representation of key stakeholders ensured a dialogue with diverse experiences rooted in local expertise and a participatory co-innovation process. These workshops helped to systematize adaptation practices, to identify how these practices are connected to functions of resilience, and to build capacity to implement on farm evaluations of resilience.

A total of sixty-four adaptation practices were identified through the workshops including, but not limited to using tolerant varieties and breeds, intercropping, silvopastoral systems, crop rotations, use of green compost, terracing, live and dead barriers, and water retention ditches. The following ten resilient functions to drought that exist in the agroecological design and management of farms were identified: (1) Reduce damage by using tolerant varieties and breeds (less sensitive, less exposure, or grown outside of the drought period); (2) Increase the vigor of the reproductive material (less stress due to exposure); (3) Increase retention of soil humidity; (4) Reduce direct sun radiation on the soil; (5) Reduce direct sun radiation on plants and productive animals; (6) Guarantee water availability for agriculture and cattle breeding; (7) Improve the physical, chemical, and biological properties of the soil; (8) Manage water rationally; (9) Regulate the microclimate; and (10) Promote physiological processes, the natural defense of plants and animals, and positive biotic interactions in the agroecosystem. These exercises helped farmers gain a greater understanding of the relationship between practices and functions for drought resilience.

Finally, as part of these workshops, the participants embarked on a process of collective construction of a methodology to evaluate resilience on farms. This process aimed to improve the skills of producers and technicians, developing their skills to better evaluate to what extent farms are resilient to drought, in order to better plan the necessary improvements in design and management. The guide included 54 indicators and covers the following sections: vulnerability (exposure and sensitivity), resistance, recovery, and transformability. Resistance functions focus on how well the exposed components (productive species, soil, water, agricultural practices, crops, and others) can resist or tolerate a weather event, while the recovery functions center on the existing ability to recover after (or even during) a weather event. The transformability functions too are important, as it is understood that adaptation to climate change must be managed as a continuous, evolving, process. In other words, it is not limited solely to the preparation before an imminent extreme event but rather involves gradually taking steps to improve the capacity of resistance and recovery (Vázquez Moreno 2017). These sections then determine the resilience index and help to identify the changes that are needed to continually increase this capacity. It is a very simple guide, designed for the self-assessment of farms by the farmers themselves or with the support of local technicians.

A study that began in 2013 in three rural livestock farms with integrated agriculture in suburban areas of Havana province (Vázquez et al. 2019) revealed that systems in agroecological transition place greater emphasis on the design and management of natural capital (soil and biodiversity, for example). However, to achieve greater resilience to drought, it is necessary to also increase physical and human capital in the areas of self-organization, infrastructure, means of production, capacity to produce enough for subsistence, and input storage capacities, among other systemic factors. It is essential that practitioners of agroecology consider the social functions that provide the capacity for transformation.

Agroecosystems must integrate design and management approaches that achieve a multifunctionality that allows for adaptation and resilience functions. Results from this research demonstrate that conventional management of agriculture is highly vulnerable to droughts and in order to build resiliency, biological diversity should be introduced to the system through the practices described above. The participatory self-evaluation constitutes the foundation to refocus strategies for agroecological transitions that prioritize adaptation to droughts and can be used as a tool for local management and innovation. The management of agroecosystems rooted in agroecological principles is essential to confront and manage the impacts of droughts caused by climate change. Agroecology is a scientific and methodological approach for the transformation of vulnerable systems to resilient systems, particularly because agroecological design and management strengthen and contribute to resistance, absorption, and recuperation functions.

CONCLUSION

This chapter briefly describes three key elements to building resilience in agriculture and food systems—reducing carbon emissions through the use of renewable energy, implementing participatory assessments for locally relevant agroecological approaches, and lifting the family farm as protagonists in transitioning agriculture and food systems for healthy humans and a healthy planet. In Cuba, what was borne out of necessity has become a genuinely viable approach because it takes a broader, socially, and politically engaged approach, which at its heart is rooted in flexibility and adaptation.

The studies discussed in this chapter have demonstrated that resilience is rooted in a deep understanding and decision-making power of the rural communities that practice family farming. However, it has also revealed that this emerging resilience needs future-proofing through investments, public policies, and educational programs that support agroecology and food sovereignty. Doing so would disproportionately contribute to national self-sufficiency and create a system of positive feedback, further reinforcing resilience. It has suggested that coherent, multisector, integrated approaches that help to add value and foster innovation allow for the formation of meaningful, lasting networks and alliances, which have far-reaching positive effects across communities.

Ultimately, agroecology does not offer a one-size-fits-all solution but rather fosters an approach that encompasses a broad range of responses that have emerged empirically to local, site-specific situations. Such an approach is not a static process but rather a continuous and evolving multi-dimensional approach that is focused on building more ecologically resilient, socially just, and economically fair farming and food systems.

NOTES

1. Credit and service cooperatives are formed by peasant farms that maintain ownership of their land; this type of cooperative works on the basis of collaboration, credit, and service management.
2. A renewable energy derived from agricultural activities to produce a solid, liquid or gaseous fuel from plant material or animal waste.

3. These institutions are: Center of Technological Applications for the Development of Guantanamo (CATEDES), the University of Oriente, the OSDE LABIOFAM companies in Guantánamo, Holguín, and Granma; the Extension, Research, and Agricultural Training Unit (UEICA), in Holguín; the Pasture and Forage Station, and the Cuban Association of Animal Production (ACPA), in Las Tunas; the Pasture and Forage Station and the Cuban Association of Agricultural and Forestry Technicians (ACTAF), in Sancti Spíritus; and the Municipal Administration Councils (CAM) in Calimete and Martí (Matanzas), Cabaiguán (Sancti Spíritus), Manatí (Las Tunas), and Urbano Noris (Holguín).

4. For the innovation of the micro-networks, the project received the Latin American Innovators Under 35 Years of Age award, granted by the MIT Technology Review in 2017.

BIBLIOGRAPHY

Altieri, M. 1995. *Agroecology: The Science of Sustainable Agriculture.* Boulder, CO: Westview Press.

Altieri, Miguel A., and Clara I. Nicholls. 2009. Cambio climático y agricultura campesina: impactos y respuestas adaptativas. March. *LEISA: Revista de Agroecología.* Accessed January 20, 2021. https://www.biopasos.com/biblioteca/CC%20y%20agricultura%20campesina%20impactos%20y%20respuestas%20adaptativas.pdf.

Casimiro Rodríguez, Leidy. 2016. "Bases metodológicas para la resiliencia socio-ecológica de fincas familiares en Cuba." PhD diss., Universidad de Antioquia. Medellín, Colombia.

Casimiro Rodríguez, Leidy, and José Antonio Casimiro. 2018. "How to Make Prosperous and Sustainable Family Farming in Cuba a Reality." *Elementa: Science of the Anthropocene* 6 (1): 77.

FAO. 2017. "El estado mundial de la agricultura y la alimentación. Aprovechar los sistemas alimentarios para lograr una transformación rural inclusiva." Organización de las Naciones Unidas para la Alimentación y la Agricultura.

FAO-HLPE. 2013. "High Level Panel of Experts on Food Security and Nutrition." Food and Agriculture Organization.

Fernandez, Margarita, Justine Williams, Galia Figueroa, Garrett Graddy Lovelace, Mario Machado, Luis Vasquez, Nilda Perez, Leidy Casimio, Graciela Romero, and Fernando Funes Aguilar. 2018. "New opportunities, new challenges: Harnessing Cuba's advances in aproecology and sustainable agriculture in the context of changing relations with the United States." *Elementa: Science of the Antropocene* 6 (1): 76.

Funes Monzote, F. R. 2018. "Integración agroecológica y soberanía energética." *Agroecología* (1): 57–66.

Giraldo, Omar Felipe, and Peter M. Rosset. 2017. "Agroecology as a Territory in Dispute: Between Institutionality and Social Movements." *Journal of Peasant Studies* 45 (3): 545–564.

GTZ. 2009. "Jatropha Reality-check, A Field Assessment of the Agronomic and Economic Viability of Jatropha and Other Oilseed Crops in Kenya." Report

Commissioned by GTZ, conducted by Endelevu Energy, World Agroforestry Centre, and Kenya Forestry Research Institute.
IPCC. 2019. "Climate Change and Land: An IPCC Special Report on climate change, desertification, land degradation, sustainable land management, food security, and greenhouse gas fluxes in terrestrial ecosystems. Summary for policymakers." Intergovernmental Panel on Climate Change.
Machín Sosa, Braulio, Adilén María Roque Jaime, Dana Rocío Ávila Lozano, and Peter Michael Rosset. 2010. *Revolución Agroecológica: El movimiento de Campesino a Campesino de la ANAP en Cuba*. Havana: Asociación Nacional de Agricultores Pequeños (ANAP).
Martín, G.J. 2019. "EEPFIH Institución que promueve los Sistemas Agroenergéticos Sustentables y Resilientes para contribuir al desarrollo territorial. X Taller de Desarrollo Local de la Universidad de La Habana." March 12–13. Havana.
Mubonderi, James. 2012. "Jatropha: The Broom of Poverty; Myth or Reality? A critical analysis of the Zimbabwean Jatropha programme in Mutoko district." Practical Action Consulting.
ONEI. 2018. "Panorama en el uso de la tierra Cuba 2017." Oficina Nacional de Información y Estadística. Havana.
ONEI. 2019. "Sector Agropecuario, indicadores seleccionados." Oficina Nacional de Información y Estadística. Havana.
Planos Gutiérrez, Eduardo Orlando, Antonio Vladimir Guevara Velazco, and Roger Rivero Vega, editors. 2013. "Impacto del Cambio Climático y Medidas de Adaptación en Cuba." Instituto de Meteorología. Havana.
Proyecto Internacional BIOMAS fase III, 2018–2021. "La Biomasa como Fuente Renovable de Energía en el medio Rural. Documento de proyecto." Havana.
Relatoría. 2019. "Taller Nacional de Jatropha curcas." April 24–26. Accessed Janury 20, 2021. https://www.youtube.com/watch?v=cKoien05oew&feature=youtu.be.
Solano O., R. Vázquez, and E. Pérez. 2000. "Evaluación de las sequías agrícolas de 1998 y 1999 en Cuba." In *Actas de la Reunión de Expertos de las Asociaciones Regionales III y IV Sobre Fenómenos Meteorológicos Adversos. Caracas, Venezuela. 12-14 de Julio 1999*, 21–29. Geneva, Switzerland: Oragnización Meteorológico Mundial (OMM).
Sotolongo, J. A., J. Suárez, C. Quezada, and G. J. Martín. 2019. "Producción integrada de alimentos y energía. La experiencia del proyecto internacional Biomas Cuba, 2019."
Suárez, J. 2019. "Producción integrada de alimentos y energía. La experiencia del proyecto internacional Biomas Cuba, 2019."
van Eijck, Janske, Edward Smeets, R. E.E. Jongschaap, Henry A. Romijn, and A.J. Balkema. 2010. "Jatropha Assessment Agronomy, Socio-Economic Issues, and Ecology." NL Agency: Ministry of Economic Affairs, Agriculture and Innovation.
Vázquez, Luis L., Janet Alfonso, Yaril Matienzo, Ana Ibis Elizondo, Deisy González, Regla C. González, Antonio Enrique Joya, Delvis Subit, Elda Consuegra, Domingo A. de León, José Raúl Martínez, María Caridad Diéguez, Adriana Pérez, Sebero Aranda, and Meybel Ríos. 2015. "Vulnerabilidad a la sequía y prácticas adaptativas innovadoras en territorios agrícolas de Cuba." *Agricultura Orgánica* 21 (3): 26–34.

Vázquez Moreno, Luis L., Yurisam Aymerich Montalvo, Amaury Díaz Rodríguez, Arisleidy Peña de la Cruz, Ramiro Cobas Durand, Elizabeth Álvarez Cambas, Loexis Rodríguez Montoya, Carlos L. García Urquía, Juan Alberto Gómez Bárzaga, Yuneisy Peña Arias, Eliecer Constante Bonanné, Yusmira Savón Vaciano, Dayanis Wilson Bott, Nielys Fonseca González, José Miguel Pérez Trejo, Carlos Fernández Louit, Rodelkys Hernández Turcás, Magalys Rodríguez Guerra, Vicente Rodríguez Oquendo, and Abady Lores Pérez. 2016. "Resiliencia a sequia sobre bases agroecológicas. Sistematización de un proceso de coinnovación participativa. Provincia de Guantánamo, Cuba." Guantánamo, Cuba: OXFAM, Cooperación Belga al Desarrollo, CITMA, and ANAP.

Vázquez Moreno, Luis L. 2017. "Funciones de resiliencia: Base para la gestión agroecológica de sistemas de producción agropecuaria expuestos a sequía." In *Nuevos caminos para reforzar la resiliencia agroecológica al cambio climático*, edited by Clara I. Nicholls and Miguel A. Altieri, 12–17. Berkeley, CA: SOCLA and REDAGRES.

Vázquez, Luis L., Aurelia Castellanos, and Vivian Leiva. 2019. "Transición agroecológica y resiliencia socioecológica a sequías en Cuba." *Celia Boletín Científico* 3: 43.

Chapter 12

The Quest for Energy Alternatives in Post-1959 Cuba

Helen Yaffe

The solar radiation that Cuba receives in one day is equivalent to the energy produced by 50 million tons of petroleum, according to Luis Bérriz, president of the Cuban Society for the Promotion of Renewable Energy Sources and Environmental Respect, known as Cubasolar. "It is greater in energy value than the petroleum Cuba consumes in five years," he adds (Bérriz 2017a). Cubasolar was set up in 1994 in the midst of Cuba's acute energy crises following the collapse of the Socialist Bloc, which saw the end of the island's oil imports. As a consequence, Bérriz (Ibid.) explains, "We were obliged to develop a new energy policy, based on our own resources."

Indeed, "simple expediency" is the usual explanation for Cuba's focus on renewable energies in the post-Soviet era, culminating in the "Energy Revolution" from 2005 onward. It is, however, an explanation that overlooks the research undertaken in this area by Cuban scientists, including Bérriz, since the late 1960s, despite the once abundant supply of cheap Soviet oil. The experimentation with renewable energies performed in research institutes and universities nationwide created the technical and institutional basis for subsequent progress when necessity prevailed. The expediency argument also fails to account for a historical quest for energy alternatives, and a commitment to democratizing access, in post-1959 Cuba based on ideological principles, not just economic factors. Both of these elements are explored in this chapter.

The Cuban Energy Revolution was motivated by multiple factors: reducing spending and dependence on oil imports and lessening Cuba's susceptibility to volatile international energy prices; the need for energy security and national defense; combatting the effects of climate change; and improving social and economic justice by raising living standards and universalizing access to energy. Creating a new energy paradigm is central to the current process of updating the Cuban economy, to the national development plan through to 2030 (*Plan 2030*), and to the government's efforts to attract foreign investors.

Nevertheless, there is a gulf between this long-standing commitment to developing renewable sources of energy in revolutionary Cuba and progress achieved to date; renewables accounted for just 4.5 percent of electricity production for the national grid in 2017 (Bérriz 2017b). In 2021, the urgency of making the transition to sustainable energy has intensified because of efforts by President Trump's administration in the United States to curtail the island's fuel imports. At the same time, U.S. sanctions restrict Cuba's access to the foreign exchange needed to finance investment in renewables.

In pre-revolutionary 1950s Cuba, 56 percent of the population had access to electricity. Nearly half the population was rural, but only 9 percent of rural dwellings had electric light, compared to 87 percent of urban homes (U.S. Department of Commerce 1956, 178, 187). U.S. investors controlled 90 percent of the island's electric services (U.S. Department of Commerce 1956, 10). In his famous 1953 trial defense speech, "History Will Absolve Me," the young lawyer Fidel Castro, who had taken up arms against the Batista dictatorship, articulated the notion of access to electricity both as a right and a commitment to develop renewable energy:

> Two million eight hundred thousand of our rural and suburban population lack electricity. . . . [The electricity monopoly] . . . extend lines as far as it is profitable and beyond that point they don't care if people have to live in darkness for the rest of their lives. The State sits back with its arms crossed and the people have neither homes nor electricity . . . today possibilities of taking electricity to the most isolated areas on the island are greater than ever. The use of nuclear energy in this field is now a reality and will greatly reduce the cost of producing electricity. (Castro 1953)

Ambitious efforts to apply science and technology to socialist development would later become a defining feature of the Revolution when it finally came to power in 1959.

THE EARLY QUEST TO DEMOCRATIZE ENERGY

"Energy is an instrument of power," asserts Bérriz. "What did the United States do when it broke off relations with Cuba? Take away our oil supply. What did the USSR do? Give us oil" (Bérriz 2017a). In August 1959, the revolutionary leadership ordered a 30 percent reduction in electricity charges and the following year it nationalized the U.S.-owned Cuban

Electric Company. Extending electricity to the entire population was imperative to reduce socioeconomic differences, especially between urban and rural Cuba (Hausman and Neufeld 2017). As minister of industries, Ernesto "Ché" Guevara told the Electrical Energy Forum of 1963 that: "Without electricity, it's impossible to locate new industrial centers and often the preferable location from other standpoints has been affected because electrical provision was insufficient for installing factories" (Guevara 1977, 135).

In October 1960, as the U.S. government introduced a partial blockade of Cuba, Guevara led a trade mission to the Socialist Bloc. He secured Soviet assistance in the electrification of the island and power plants were imported from the Soviet Union, Czechoslovakia, East Germany, and France. However, because the Cuban electrical system operated with 60 hertz, and the eastern European socialist countries used 50 hertz, the equipment had to be adapted by technicians before being sent to Cuba. Another fundamental problem was the lack of technical knowledge on the island (Sáenz 2005).

The agreement to export Cuban sugar to the USSR and import Soviet oil at below world market prices inevitably perpetuated the island's oil import dependence and structural underdevelopment. This relationship also fostered the increasing mechanization of agriculture, as the USSR supplied the island with diesel-fueled tractors, petroleum-based pesticides and fertilizers. Incentives were lacking for the development of alternative energy sources. Nonetheless, while energy specialists were trained to replace the technicians leaving the island and to staff the extended electrical provision in the 1960s–1970s, a search for alternatives to hydrocarbons had begun. Table 12.1 tracks the key areas of research and investigation across the country from the 1960s to 1980s.

According to Bérriz (2017a), by the late 1980s, "from the scientific perspective, Cuba came to be a power in the development of renewable energy sources. It had its first building completely supplied with renewable sources of energy . . . and was one of the most advanced in renewable sources in tropical conditions." Nonetheless, beyond experimentation, the application of these technologies was very limited. This was partly due to the lack of technology and investment, as well as access to them, as a result of the U.S. blockade. Perhaps more decisive, however, was the absence of commitment or incentive to restructure the energy mix away from hydrocarbon fossil fuels, given that the island's industry and agriculture were orientated toward cheap imported oil from the USSR. Indeed, Cuba had committed to the Green Revolution, a system that requires the massive use of fossil fuels in the form of natural gas-based fertilizers, oil-based pesticides, and diesel fuel for tractors and other farm-based machinery. Cuba had 90,000 Soviet tractors in the 1980s and the island's agriculture, "was more industrialized than any other

Table 12.1 1960s and 1980s: Development of Key Areas of Research (Bérriz n.d.; Bérriz 2017a; Curbelo et al. n.d.)

When	*What Happened*
1968	Physics Faculty, University of Havana, established a course on photovoltaic devices. By 1975, it had produced the first Cuban photovoltaic solar cell made with crystalline silicon, the standard first-generation panel material.
1970	Urban population entirely connected to the National Electroenergetic System (SEN), providing electricity at 110 kV. During the decade, the SEN was interconnected at voltages of 220 kV.
1970s	Scientists in the sugar industry worked to develop renewable energy from *bagasse,* improving the efficiency of furnaces and boilers in sugar centrals where *bagasse* had long been used as fire fuel.
1975	Headed by Luis Bérriz, the *Grupo de Energía Solar* (Gensolar) was set up to develop renewable energies. Directed by the new *Principal State Program of Research into Use of Solar Energy in Cuba*, approved by the First Congress of the Cuban Communist Party, it developed the first Cuban solar heaters, solar dryers, distillers, sea water purifiers, concentrators, and technologies for the use of solar energy in the cultivation of microalgae.
Late 1970s	Group of Technical Assistance on Energy created within the Ministry of Basic Industry to develop strategies for energy saving and developing renewable energies. In 1981, their *Report on Cogeneration* promoted the use of renewable energy in the sugar industry and other areas.
Early 1980s	Physics Faculty, University of Havana, improved the Cuban-made solar cells with the use of gallium arsenide, increasing efficiency of solar panels. First Cuban solar heaters produced.
1983	Environmental Physics Group set up in the Ministry of Construction to promote a culture of "passive" solar energy use, bioclimatic architecture, and regulatory norms for efficient use of energy in buildings. In addition, the National Energy Commission (Comisión Nacional de Energía, or CNE) was created to rationalize energy consumption and increase use of national resources, especially renewable sources. Installed 200 (mini & micro) hydroelectric power stations to supply 34,000 people in mountainous regions. Worked to expand the use of renewable energies, mainly hydroelectricity, solar thermal, wind power, and biogas.
1984	Center for Research into Solar Energy (Centro de Investigaciones de Energía Solar, or CIES) set up in Santiago de Cuba, to develop renewable energy sources.
1984	National Institute of Hydraulic Resources founded the Biogas Group Villa Clara.
1986	Ministry of Communications laboratory began assembling photovoltaic solar modules made of monocrystalline silicon with an annual capacity of 200 kWh.
1988	Combustion Laboratory was established in Cienfuegos to research combustion of solid renewable fuels.

Latin American country and exceeded the US in its use of fertilizer" (Arthur Morgan Institute for Community Solutions 2006).

Even so, the fact that Cuban scientists conducting research in multiple institutions, working under government directives to develop renewable energies since the 1960s, facilitated the urgent uptake of new practices when the shock of the Soviet collapse made doing so an obligation.

THE EXPEDIENCY OF THE SPECIAL PERIOD: AN ACUTE ENERGY CRISIS

The Soviet Union's unilateral decision in 1990 to reduce oil exports to Cuba presaged the so-called "Special Period" on the island. From 13 million tons in 1989, Soviet oil exports to Cuba plummeted to 1 million tons in 1992 and zero thereafter (Rodríguez 2016). Cuba was crippled by critical scarcities of hydrocarbon energy resources, fertilizers, food, medicines, cement, spare parts, equipment, and resources in every sector. Power outages became a defining feature while the island's transport system ground almost to a halt.

In terms of agriculture, tractors were replaced by human and animal labor, while organic fertilizers, crop rotation techniques, and urban gardens were developed as people used vacant plots and began to plant fruit and vegetables (Arthur Morgan Institute for Community Solutions 2006). The army was sent to farm fallow land, nearly 40 percent of state farms were converted into cooperatives and non-state farmers' markets were opened. In small towns and cities, people turned to horses for transportation. The 1950s Cadillacs and Chevrolets were parked up alongside Russian *Moskvitch* to rust in garages while 1.2 million bicycles were imported from China and half a million more were produced domestically (Arthur Morgan Institute for Community Solutions 2006). New centers, ministries, and programs were established nationwide in the 1990s to pursue alternative energies with greater urgency, as summarized in table 12.2.

Despite these efforts, Bérriz admits, "the country's energy policy remained basically oil-fueled, with centralized electricity generation in a few thermoelectric plants" Bérriz (2017b, 7). The principal difference in the post-Soviet 1990s was that in the absence of imported oil, Cuba was burning domestic crude oil reserves. Many Cuban policy makers still believed that renewable energy sources were scarce, unstable, expensive, inefficient, inadequate for meeting Cuba's energy needs and would require extensive territorial use. Other obstacles were the cost of solar panels produced in Europe, until mass production in China saw costs plummet, and the need for technology to enable renewable-generated energies to be fed into the national electricity grid.

Table 12.2 New Centers, Ministries, and Programs Established in the 1990s to Pursue Alternative Energies (Arthur Morgan Institute for Community Solutions 2006; Nachmany et al. 2015)

When	*What was Established*
1992	Center for the Study of Renewable Energy Resources set up as a university teaching-research center within the José A. Echeverría Polytechnic Higher Institute (known by the Spnaish acronym, ISPJAE) in Havana.
1992	Center for the study of Sugar Thermoenergy founded in the Central University of Las Villas, for experimentation related to sugar mill energy production.
1993	National Energy Sources Development Program approved by the National Assembly of People's Power to improve energy-efficiency and renewable energy use.
1994	New Ministry of Science, Technology, and the Environment (Ministerio de Ciencias, Technología y Medio Ambiente, or CITMA) established.
1994	Hydro-energy Research and Development Area set up in Villa Clara to coordinate researchers and technicians advising government bodies on hydropower, build hydro-energy facilities, and train specialists.
1994	Cubasolar founded to promote all forms of solar energy: biomass, biogas, hydropower, sea and wind, solar photovoltaic, solar thermal and passive use.
Mid-1990s	Integrated Center for Appropriate Technology founded within the National Institute of Hydraulic Resources to develop technologies to supply water to the population with windmills, hydraulic rams, rope pumps, and winches.
1996	Center for the Study of Efficient Energy founded to develop technologies for the combustion of biogas, mainly bagasse, and the use of biogas in internal combustion engines.
Late 1990s	Solar heaters factory built in Morón city, Central Cuba.
1997	CITMA creates the Group of Solar Energy Technological Applications (known by Spanish acronym, GATES) in Guantánamo to install photovoltaic systems in areas beyond the national grid.
1997	Ministry of Education (Ministerio de Educación, or MINED) launched an Energy Saving Program (PAEME), "to involve citizens in the effort to save energy" (Nachmany et al. 2015, 6). Energy Festivals educate Cubans about energy efficiency and conservation, and schools incorporate the theme in physics, economics, and environmental courses.
1999	Cuba's first wind farm inaugurated on Turiguanó island, doubling as a center for wind energy studies.
2000 onward	Cubasolar and Ecosol Solar help form provincial groups of specialists in photovoltaic technologies.

There was, however, commitment to developing renewable energies among Cuba's top leadership. A leading member of the Cuban government and the Cuban Communist Party (Partido Comunista de Cuba, or PCC), Vilma Espín had graduated in chemical engineering in the 1950s, before joining Cuba's revolutionary movement.[1] In the early 1990s, she requested a meeting with Bérriz. "I prepared to convince Vilma of the great importance of using renewable energy sources," recalled Bérriz, "but when we sat down to talk, Vilma started to convince me of the great importance of using renewable energy sources! I kept quiet and thought, *I don't have to do anything here*" (Bérriz 2017a).

By 1994, Cuba's electricity crisis was critical. The lack of fuel meant that generation capacity was 40 percent below its potential (Ibid.). Industries were paralyzed while energy shortages were rationed; across the island, residents experienced regular rotating blackouts lasting 8–12 hours, without lights, air fans, televisions, and fridges. Cubans suffered long waits for a significantly depleted stock of public transport to return from work or college. On alternate days, they could arrive in a darkened neighborhood, obliged to cook, wash, clean, or do homework by candle or kerosene flame.

RENEWABLE ENERGY: FUELING THE BATTLE OF IDEAS

In December 1999, Cuba launched the "Battle of Ideas," initially as a campaign to demand the return from Miami of the shipwrecked Cuban boy Elián González to this father in Cuba. It expanded to incorporate ambitious socioeconomic and educational programs, with youth as the principal protagonists and beneficiaries. Many of these programs required the installation of solar panels in remote areas, facilitating access to televisions, videos, and computers. In December 2000, an electronics factory in Pinar del Río began to manufacture Cuban solar panels and within a year, 2,000 schools beyond the reach of the national grid had been supplied with them, including those in isolated, mountainous areas.[2]

In 2002, the Council of Ministers set up the Renewable Energy Front to advocate policies and devise strategies to strengthen the sustainable use of renewable energies. The previous year, the Center for Information Management and Energy Development, known as Cubaenergía, had been created through the merging of several research and technology institutes. Its eighty scientists and technicians focused on energy efficiency, renewable energies, and providing scientific-technical services to state enterprises and government ministries at a provincial level. In 2004, Cubaenergía launched a pilot project, with United Nations funding, to increase the contribution of renewable energies on the Island of Youth (Isla de la Juventud).[3] They identified biomass, produced as a residual from the woodland, which covers over 54 percent of the Island, as a potential source of electricity generation and designed demonstration plants (Curbelo Alonso 2017). They also built a small wind farm with folding towers that can be lowered during hurricanes.

Starting in 2000, the trade and cooperation agreements signed between Cuba and Venezuela saw tens of thousands of Cuban medical professionals and educators pour into Venezuela's poor and marginalized neighborhoods, while Cuba received up to 90,000 barrels of Venezuelan crude oil a day (Azicri 2009). The deal bought Cubans some respite from the blackouts and bus queues of the special period and gave many millions of Venezuelans access to quality medical care. However, the "oil for doctors" program did not solve Cuba's energy problems. In 2003, Cuba's aged thermoelectric power stations were converted to burn the low-quality, sulfur-heavy oil extracted from the island's shallow coastal waters. A Cuban expert lamented: "Without the embargo this high sulfur content oil would have been used for making asphalt, not for energy production. But after the Soviet support was gone we had to turn to our national oil reserves" (quoted anonymously in Käkönen et al. 2014, 11). The result was a series of breakdowns, starting in 2004 with the major thermoelectric generator in Matanzas. This effectively set Cuba back to the acute energy crisis of 1994, with electricity generation down to 38 percent of its potential (Bérriz 2017a). Blackouts returned, this time resulting from maintenance work, and the withdrawal of the faulty thermoelectric plants in Matanzas and Camagüey.

These problems were compounded by hurricanes knocking out parts of Cuba's electricity supply: Pinar del Río lost electricity for a fortnight and Trinidad de Cuba for a month after Hurricane Dennis struck as a Category 4 storm, unseasonably early in July 2005. Power cuts and communication systems broke down as 1,025 electricity posts were felled, 21 municipalities were left without electricity and 2.5 million people without a direct water supply (Yaffe 2005).

"Look how weak we were!" exclaims Bérriz, "If anyone had thrown a bomb at Guiteras [power station in Holguín] you would have wiped out half of our electro-energetic system. If they had thrown seven, they would have finished us off!" Essentially, Cuba's electricity supply was generated through seven power stations, meaning the island was extremely vulnerable to attack. Between the military invasion of Afghanistan in 2001 and Iraq in 2003, the U.S. undersecretary of state, John Bolton, had added Cuba to the list of "rogue states" and "state sponsors of terrorism"; military aggression certainly seemed possible to the Cubans. "It is not just the lights that go off," points out Bérriz, "it wipes out industrial production, the warehouses, food, everything." Furthermore, he adds, "we can no longer live without electricity, we are too accustomed to it" (Bérriz 2017a). A solution to Cuba's vulnerability was sought through the Energy Revolution.

2006: "YEAR OF THE ENERGY REVOLUTION"

The Energy Revolution was a major state initiative to improve energy security and efficiency. It saw efficient new power generators installed in a "distributed" system, emphasis on renewable energy increased, a progressive electricity tariff introduced, and old durable goods replaced with energy-saving equipment. The intention was to increase Cuba's capacity for electricity generation to four times its needs, consigning blackouts to the past and addressing the island's energy dependence.

During a seminal speech to student leaders at the University of Havana in November 2005, Fidel Castro framed the question of energy (cost to the state, cost to the consumer, state subsidies, theft and wastage, energy efficiency, and conservation) in relation to the Battle of Ideas and to the survival of the Cuban Revolution. State subsidy for fuel was so high that bills were negligible, he said. "No one knows the cost of electricity, no one knows the cost of petrol, no one knows its market value" (Castro 2005). This placed an immense burden on the state budget and was a disincentive to individuals and businesses to conserve energy. "We discovered that a *paladar* [private] restaurant consumed 11,000 kilowatts and that this stupid State was subsidizing the owner . . . [with] more than 1,000 dollars a month," he complained (Castro 2005). The Energy Revolution would see the state reduce its energy subsidy, inducing awareness of consumption and saving among the Cuban people. Fidel Castro estimated that Cuba's new energy policy could save two-thirds of Cuba's existing energy consumption, more than $1.5 billion a year, or 25 billion Cuban pesos, nearly double the sum of Cuban wages.

The transition to a decentralized, distributed system of energy generation sought to reduce Cuba's vulnerability to natural disasters and external aggression. The national security motivation has been overlooked in external reports that focus on three other drivers: the decline of oil supplies, extreme weather conditions due to climate change, and economic crises (Käkönen et al. 2014; Seifried 2013). 1,531 small generators with a total output of 3,072 MW were imported from Danish and Spanish companies and installed throughout the island (Curbelo 2017; Seifried 2013).[4] Fidel Castro was personally involved, says Curbelo: "Fidel invited the presidents of those businesses to Cuba and negotiated the price, everything . . . we sought practically the best companies in the world; obviously they couldn't be American because of the US blockade" (Curbelo 2017). This decisively reduced Cuba's vulnerability, confirms Bérriz: "Now we have thousands of electric plant generators.

The more electricity generators we have the stronger we become" (Bérriz 2017a). In addition, the distributed system is conceptually closer to the kind of localized generation required for a switch to renewable energies (Curbelo 2017). Thus, while solving the energy crises of 2003–2005 and improving the island's energy security, the Energy Revolution was also a step toward a "new energy paradigm" in which renewable energies would play an increasing role (Arrastía Avila 2008).

In 2005, blackouts lasting more than one hour had occurred on at least 224 days, but by 2007, power outages caused by the lack of generation capacity had been eliminated (Seifried 2013). In the same period, Cuba reduced its carbon dioxide emissions by some 5 million tons; equivalent to 20 percent of the island's total emissions in 2002 (Arrastía Avila 2008). Cuba was also second in the world, after Denmark, in terms of distributed energy production (Käkönen et al. 2014). In addition, more than 4,000 emergency back-up systems had been installed in hospitals, food production centers, schools, and other critical sites, representing 500 MW of emergency back-up power (Guevara-Stone 2009). Upgrades were made to 120,000 electrical posts, nearly 3,000 kilometers of cable was laid and half-a-million electrical meters were installed, reducing wastage and the amount of oil needed to generate 1 kWh of electricity from 280 grams in 2005 to 271 in 2007. "While this might seem like a small saving, it translates into thousands of tonnes of imported oil annually," wrote Laurie Guevara-Stone, adding that "in 2006–2007 Cuba saved over 961,000 tonnes of imported oil through energy saving measures" (Ibid.).

IMPROVING ENERGY EFFICIENCY AT HOME

For most Cubans, the first sign of the Energy Revolution began in July 2005 when teenage social workers knocked at their doors delivering free energy-saving fluorescent bulbs. Within six months, around 9.4 million incandescent bulbs had been replaced, making Cuba the first country in the world to complete the switch (Suárez et al. 2012). Conservative estimates are that this generated an annual saving of 354 million kWh, 3–4 percent of total electricity consumption (Arrastía Avila 2008; Käkönen et al. 2014; Seifried 2013). In state enterprises, 800,000 inefficient fluorescent tubes with magnetic ballasts were exchanged for more efficient lamps with electronic ballasts (Seifried 2013).[5]

The social workers drew up inventories of each household's electrical equipment for the subsequent campaign to replace inefficient and aging domestic appliances with more efficient substitutes imported from China. A Finnish report on the Energy Revolution noted: "The biggest 'energy monsters' were the old refrigerators from the 50s and the ventilators [air fans] that have been made from old Russian washing machines (a remarkable 'grass-roots' innovation in itself of course, but a very energy consuming one)" (Käkönen et al. 2014, 14). A similar German report calculated that each new Chinese refrigerator brought an annual saving of about 450 kWh per unit (Seifried 2013, 8). With over two-and-a-half million old fridges replaced, this totaled 1,148 million kWh; an annual sum of "about 230 million euros saved in operating costs (mainly fuel costs)" (Ibid.). Also replaced were over one million air fans, 265,500 air conditioners and ventilators, 230,500 televisions, and over 268,000 water pumps. Simple electric hotplates were distributed, along with 3.5 million rice cookers and 5.5 million pressure cookers (Käkönen et al. 2014, 14). According to the German report: "While the vast majority of households cooked with kerosene and LPG [liquid petroleum gas] until early 2006, within a few months some three million households were converted almost completely to electric cooking. Gas stoves remained only in areas where there was a gas supply" (Seifried 2013, 8–9).

The purchase and replacement of this energy-saving equipment required a huge investment by the state. Energy-saving bulbs and fans were distributed to households free of charge, but Cubans were charged for the other items. A credit scheme was set up for most of the population who could not pay upfront (Ibid.). Within two years, Cuban banks financed over 4.5 million household appliances to a value of 9 billion Cuban pesos (*Juventud Rebelde* 2007). Repayments were adjusted to household incomes. In 2005, the average individual's monthly salary in Cuba was 330 Cuban pesos (CUP) (Centro de Estudios de Población y Desarrollo 2010). Households with a monthly income of up to 225 pesos were offered credit at an annual interest rate of 2 percent and a repayment period of 10 years. The interest rose to 3 percent for households with incomes between 226 and 450 pesos; to 4 percent for incomes between 451 and 600 pesos, and so on. The higher the income, the greater the rate of interest. "With an income of 1801 CUP and more, loans were no longer granted," explained Dieter Seifried (2013, 16), author of the German report. 92 percent of Cubans who requested credit paid between 10 and 75 pesos monthly (*Juventud Rebelde* 2007).

How much the equipment cost the Cuban government has not been revealed, but based on his own estimations plus the fuel costs saved, Seifried calculated that it was amortized in less than two years, concluding that "we obtain a benefit-cost ratio for the Cuban economy of about 10 [to one]" (Seifried 2013, 8, 11).[6] The switch to electric cooking alongside increased ownership of consumer durables saw electricity consumption increase by 13 percent. Despite this, these measures saved Cuba about 250,000 tons of oil equivalents annually (Seifried 2013). In May 2006, Fidel Castro told the state electric company: "We are not waiting for fuel to fall from the sky, because we have discovered, fortunately, something much more important—energy conservation, which is like finding a great oil deposit" (Guevara-Stone 2009, 3).

A new progressive electricity tariff reduced the state subsidy and encouraged saving, without hurting the country's poorest people; where consumption tripled, the cost of electricity per unit increased nearly fifteen times. In 2011, the top rates were revised upward, so the highest band was over fifty times that of the lowest, with no change for low or medium consumption households (Seifried 2013, 13).

These energy-efficiency measures were largely directed at the residential sector for two reasons. First, because it accounts for some 45 percent of Cuba's electricity consumption, an unusually high proportion, which reflects low levels of industrial activity (Curbelo 2017). Second, the investment required was less than that of replacing capital goods, equipment, machinery, infrastructure, and transport. "The last industrial plants introduced into the country are from the 1980s and they were built to use fossil fuels and without prioritizing energy efficiency," explains Curbelo. Indeed, modern plants might produce the same product with half the energy consumption. The same applies to transport. "Energy efficiency measures in transportation are associated with very high investment costs and we would practically have to renovate the entire automotive fleet," notes Curbelo (Ibid.).

EMBEDDING THE ENERGY REVOLUTION

"Educating a population of 11.5 million about energy is a tall order," pointed out an article on the role of education in the Energy Revolution. "Cuba's energy education program focused on creating a new energy culture and on achieving sustainable development" (Arrastía Avila Guevara-Stone 2009, 31). Notably, MINED's energy-saving program, PAEME, took the Energy Revolution into schools and communities. Hundreds of educational "festivals" were held, and thousands of thematic workshops organized (Guevara-Stone 2009). Between 2006 and 2008, there were 22,000 television and radio broadcasts, and 1,600 newspaper articles about the measures underway, and more than 1,100 discussions took place in communities (Seifried 2013).

New institutions were set up within the framework of the Energy Revolution, including the National Group for Renewable Energy Sources, Energy Efficiency and Cogeneration, which had fourteen commissions covering different types of renewable sources of energy and efficiency, coordinated by the government, to research and propose ways of developing Cuba's renewable energy potential. A new Vice Ministry for Renewable Energies was created in the Ministry of Basic Industry (Nachmany et al. 2015).

National programs were launched to generate electricity from wind energy, solar water-heating for domestic, social, and industrial purposes, to develop hydro and solid waste energy capacities and to research geothermal, ocean energy, and other technologies (Avila 2009; Käkönen et al. 2014).

By 2009, 6–7 percent of the Cuban people were receiving electricity from gensets, small hydropower stations, or solar panels in off-grid areas. Total installed electricity production capacity was some 6,000 MW, while energy demand in Havana was around 500 MW. Electricity production capacity was 42 percent from the distributed systems (Käkönen et al. 2014). The contribution of renewable energies to *electricity* production was 4 percent (in 2011), but it was 20 percent in primary *energy* production, given the use of bagasse-fired energy in the sugar industry. Hence the Finnish report concluded that: "The emphasis on increasing the use of renewable energy sources in the overall energy mix has, however, been slow and is not yet clearly visible in the statistics" (Käkönen et al. 2014, 16).

In 2010, Cubaenergía's renewable energies project on the Island of Youth passed from the experimental to the investment phase and the Ministry of Energy and Mines (Ministerio de Energía y Minas, or MINEM) took control (Curbelo 2017). Between January and September 2014, diesel use on the Island of Youth was reduced by 329 tons, saving the equivalent of $253,000 on the international market (*Revista Eólica y de Vehículo Eléctrico* 2014). In 2015, two solar parks contributed 1,039 MW to the electro-energetic system, equivalent to some 900 tons of fuel, saving $517,621 in oil imports for generating electricity. A third solar park was under construction.

Since 2015, as part of a new project with the European Union, worth over €3 million, Cubaenergía has worked to demonstrate the benefits of biogas, photovoltaic solar panels, and solar dryers for agriculture. The organization is also developing a "bioenergy technology transfer" project in Matanzas province. "Bioenergy" refers to the use of biogas, wood waste, forest biomass, and the production of biodiesel from a plant that grows well with little need for water, with the scientific name *Jatropha Curcas*, known as *Piñón de botija* in Cuba. The project fosters these technologies and the equipment that uses them. "Cubaenergía is responsible for showing the country that the potential exists, and what kind of regulatory measures would have to be taken to achieve its potential use," states Curbelo (2017).

A BRIGHT AND RENEWABLE CUBA?

According to Bérriz, in contemporary Cuba, the pioneer in renewable energy use in the island is the Revolutionary Armed Forces (Fuerzas Armadas Revolucionarias, or FAR), headed by Raúl Castro from 1959 until 2008.

"Most military buildings have solar heaters and they have many solar parks. They have worked a lot with wood dryers and solar dryers for their agricultural products" Bérriz (2017a) contends. For the FAR, self-sufficiency is an issue of national security, and this motivates the military's engagement in productive and economic activities. Renewable energies and energy efficiency bring Cuba closer to energy sovereignty.

Despite the progress made and the aspirations of its proponents, by mid-2017, Cuba's energy structure remained highly dependent on imported fossil fuels, with the associated economic and strategic risks that implied. Renewable energy sources contributed just 4.5 percent of electricity production for the national grid, compared to 45 percent crude oil, 18 percent motor oil, 15.1 percent thermal fuel, 14.1 percent natural gas, and 3.3 percent diesel. The national development plan seeks to increase the share of renewable energy sources to 24 percent of electricity production by 2030 (Bérriz 2017b).

These statistics, however, fail to capture fully the extent of renewable energy use in Cuba, because they refer to the production of electricity in the national grid, not the production of energy *per se*. This distinction is overlooked even in Cuban reports, which refer to the aim of raising renewable energy sources to 24 percent of energy production by 2030. "We already have 30 percent of the country's energy produced by renewable energy sources!" exclaims Bérriz. "All the electricity produced by the sugar refineries for their own consumption is not included in the current statistic citing of 4 percent." Another example is the aqueduct at Baracoa in eastern Cuba; it was a huge consumer of electricity until it was reconstructed to use gravity, instead of an electrical pump, to supply water. "Before it used so much electricity, and now it doesn't use any, so it no longer appears in the statistics," points out Bérriz (2017a).

A second statistical error is not factoring in rising production. Reports often describe the stated aim as a six-fold increase in the contribution of renewable energy sources to electricity production (from 4 percent to 24 percent), when in fact it represents a multiplication of ten, because electricity production is expected to increase from 18 terawatts (TW) to 30 TW in 2030. Bérriz (Ibid.) explains: "4 percent of 18 is 0.72 and 24 percent of 30 is 7.2; so the amount of electricity that will be produced for the national grid will be a multiplication by 10." For example, Table 12.3 illustrates the stock of renewable energy as of 2017.

There had also been progress in energy efficiency. By mid-2017, 2 million fluorescent bulbs had been replaced with LED bulbs in the residential sector, with another 11 million to go, plus another 250,000 public lights. Over half a million induction cookers had been sold to Cuban households, with the goal being to replace 2 million electric resistance cookers. 550 square meters of solar heaters had been installed in Cuban homes, out of the planned 200,000

Table 12.3 The Existing Stock of Renewable Energy Source, 2017. Data from the Office of Renewable Energies, Ministry of Energy and Mines (Ministerio de Energía y Minas, or MINEM) (Bérriz 2017b)

Technology	*Installed capacity*	*Installations*
Sugar cane biomass	470 MW	57 sugar mills
Wind	11.1 MW	4 wind parks
Solar Photovoltaic	37 MW	22 photovoltaic solar parks
	5543 MW	Rural schools
	467 MW	Doctors surgeries
	1882	Rural TV centers
	2500 MW	149 isolated houses
Hydroenergy	66 MW	149 Hydroelectric plants
Biogas	0.82 MW	5 biogas plants
Non-sugar biogas	0.5 MW	4 biogas plants

square meters, reducing the electricity consumed in those homes by an average of 12 percent. By late 2018, 31,000 facilities in Cuba were using solar heaters, 9,476 had solar panels, 9,343 had windmills, 3,234 had biogas plants, and there were 147 hydroelectric facilities and 22 solar farms (MINCEX n.d., 79).

"Natural resources and the environment" is one of six strategic areas in the *Plan 2030*. Among the twenty-one "specific objectives" outlined is that of increasing energy efficiency and the development of renewable energy sources, "which, among other benefits contributes to reducing the generation of greenhouse gases, reduces climate change and promotes less carbon-intensive economic development" (PCC 2017, 20). A $4 billion investment will be required to install 2144 MW of new electrical power by 2030, to be generated via nineteen bioelectric plants, fourteen wind parks, solar parks producing 700 MW, and seventy-four small hydroelectric stations. The plan involves extending the use of solar thermal energy, forest biomass, solid urban waste, agricultural and organic industrial waste (Bérriz 2017b). 500 industrial biogas plants are to be constructed to treat waste and produce energy. Similar plants will tackle waste contamination from farms: 7,000 biogas plants using pig excrement and another 1,700 for cow excrement. Ambitious plans for forest biomass include converting *marabú*, the leguminous tree with deep roots which spread like a weed over 2 million hectares of uncultivated agricultural land, about 18 percent of the country's territory, from a curse into a resource. An article in *The Economist* about a British–Cuban joint venture in this area described it as "Cuba's wonderful weed," explaining that: "Three tonnes of the stuff can produce as much electricity as a tonne of fuel oil" (*The Economist* 2017). 67 sawmills will also have small biomass electricity generators. The first of these are under development.

By 2017, three Cuban factories in Morón, Ciego de Ávila, and Cayo Coco were manufacturing solar water heaters. "We would like to have a factory in every province. And it is one of the things which we are promoting," says Bérriz, who is committed to converting Cuban homes from net consumers of energy into net producers, or "positive energy homes." It is possible, he insists, not a utopia, if the installation of solar panels is accompanied by moves to cook with biomass and heat water with solar heaters (Bérriz 2017a, 2017b). In Cuba, residential use accounts for nearly half of total electricity consumption and half of that is used for cooking food and for heating water for bathing. "Ninety percent of the Cuban population heats water to bathe themselves. In summer, people heat water to wash!" exclaims Bérriz. "If we put solar heaters in the houses, this immense quantity of electricity will no longer be consumed. Instead of using petroleum to heat water, we could use the sun. The technologies exist, what we have to do is use them" (Bérriz 2017a). In August 2019, a new program was approved to sell solar heaters to residents in Havana (*Cubadebate* 2019). Lights, televisions, and other equipment use a much smaller proportion of the electricity consumed. Implementing the new "Development of Renewable Sources and Efficient Use of Energy" law, approved on March 23, 2017, will be an important step in this direction.

THE ROLE OF FOREIGN INVESTMENT

Cuba's ability to achieve its targets is dependent on the government's capacity to attract foreign investment, including Economic Partnership Agreements. The 2016/17 *Portfolio of Foreign Investment* included $3 billion sought for renewable energy projects, with the projected potential to add 2.1 gigawatts (GW) of capacity from wind, solar, biogas, and biomass plants (Morales 2017).[7] The 2017/18 *Portfolio* listed thirteen projects for renewable energy development (MINCEX 2017).

The Cuban government has already contracted Spanish renewable energy corporation Gamesa to build seven wind farms in eastern Cuba with a total generating power of 750 MW. These will be added to four existing wind parks in Holguín, Ciego de Ávila, and the Island of Youth. Another plan is to generate 440 MW of power from solar energy by 2020. Cuba has received a $15 million loan from the Abu Dhabi Development Fund to develop four 10 MW solar power plants using photovoltaic silicon panels (Morales 2017). Given the urgency in securing foreign investment in these areas, the Cuban government is inviting 100 percent foreign-owned investments for these projects, an ownership status rarely permitted in revolutionary Cuba.

China has been a key partner, offering training for Cuban technicians, technology, and credit to help Cuba augment its domestic production of solar panels. Some 60,000 Cuban panels were manufactured in 2016, all of them with Chinese equipment, and the plan is to increase production threefold and install more automated machines (Voss 2016). Chinese capital stepped in to assist a British company, Havana Energy, which entered a joint venture with Cuban state company Azcuba in 2010 to develop biomass plants using Cuban bagasse and *marabú*. The British partners struggled to access financing from Europe, because of the U.S. blockade, so Shanghai Electric stepped in as the major shareholder (Voss 2016). By early 2018, two biomass plants were under construction, which according to a business report, would "contribute 4 percent more to Cuba's renewable energy share" and were expected to foster significant future foreign investment (*New Energy Events* 2018).

CONCLUSION

The tightening of the U.S. blockade by the Trump administration has threatened to scupper these plans. In May 2019, Title III of the 1996 Helms–Burton Act was fully enacted, after being suspended for twenty-three years. It authorizes U.S. nationals with claims to property nationalized in Cuba after the Revolution to file suits in U.S. courts against persons "trafficking" in that property. Its principal purpose is to scare foreign investors away from Cuba, by raising the perception of risk and uncertainty. Meanwhile, threatened with fines levied against them by the U.S. Treasury's Office for Foreign Asset Controls, shipping companies have stopped delivering Venezuelan oil to Cuba, despite the illegality of the extraterritorial imposition of the U.S. blockade. The need to replace hydrocarbons with renewable energies in Cuba is thus greater than ever. At the same time, the Cubans' ability to access the international finances needed to invest in the new technologies is also obstructed by the U.S. blockade. This creates a uniquely complex situation for revolutionary Cuba. The quest for energy alternatives must go on.

NOTES

1. In 1960, Vilma Espín founded the Federation of Cuban Women (Federación de Mujeres Cubanas) (FMC) and served its president until her death in 2007.
2. Subsequently, the factory began to produce solar cells, the building blocks of solar panels.
3. The Isla de la Juventud, 50 kilometers South of mainland Cuba, is the 7th largest island in the Caribbean, with a population over 86,000.
4. Seifried's (2013, 19) report adds: "improvements in the transmission and distribution networks were made to reduce the transmission and distribution losses."

5. As a result of the low quality of energy saving bulbs imported from China, the Cubans began producing and distributing LED bulbs (Curbelo 2017).

6. Seifried (2013, 11) adds: "This exceptionally large advantage in favour of energy efficiency is partly due to the low efficiency standard before the Energy Revolution, and also to the high cost of power generation in the Cuban energy sector."

7. The 2016–2017 portfolio of opportunities for foreign investment included 123 projects in energy and mining, nearly one-third of the total (Morales 2017).

BIBLIOGRAPHY

Arrastía Avila, Mario Alberto. 2008. "Distributed Generation in Cuba: Part of a Transition Towards a New Energy Paradigm." *Cogeneration and On-Site Power Production* November-December.

Arrastía Avila, Mario Alberto, and Laurie Guevara-Stone. 2009. "Teaching Cuba's Energy Revolution." *Solar Today* January/February: 30–33.

Arthur Morgan Institute for Community Solutions. 2006. *The Power of Community: How Cuba Survived Peak Oil.* Documentary.

Azicri, Max. 2009. "The Castro-Chávez Alliance." *Latin American Perspectives* 36 (1): 99–110.

Bérriz, Luis. n.d. "Algunos aspectos del desarrollo histórico del uso de las fuentes renovables de energía en cuba." Document. Havana.

Bérriz, Luis. 2017a. Interview with Helen Yaffe. July 5. Havana.

Bérriz, Luis. 2017b. "La Política Energética Cubana en el Camino del Desarrollo Sostenible." Presentation. June. Havana.

Castro, Fidel. 1953. "History Will Absolve Me." Accessed January 20, 2021. https://www.marxists.org/history/cuba/archive/castro/1953/10/16.htm.

Castro, Fidel. 2005. "Speech delivered by Dr. Fidel Castro Ruz, President of the Republic of Cuba, at the Commemoration of the 60th Anniversary of his admission to University of Havana, in the Aula Magna of the University of Havana." November 17. Accessed January 20, 2021. http://www.cuba.cu/gobierno/discursos/2005/ing/f171105i.html.

Centro de Estudios de Población y Desarrollo. 2010. *Salario En Cifras: Cuba 2010.* Havana: Oficina Nacional de Estadísticas.

Cubadebate. 2019. "La Habana: Venderán calentadores solares de agua para el sector residencial." August 27. Accessed January 20, 2021.

Curbelo Alonso, Alfredo. 2017. Interview with Helen Yaffe. July 5. Havana.

Curbelo Alonso, Alfredo, Bárbara Garea Moreda, and Antonio Valdés Delgado. n.d. "Generación de electricidad a partir del bagazo en Cuba." Accessed January 12, 2021. http://www.fao.org/docrep/t2363s/t2363s0n.htm.

Guevara, Ernesto "Ché." 1977. "Discurso Clausura del Fórum de Energía Eléctrica." November 20, 1963. In *Escritos y Discursos*, Vol. 7, 135. Havana: Editorial de Ciencia Sociales.

Guevara-Stone, Laurie. 2009. "La Revolución Energética: Cuba's Energy Revolution." *Renewable Energy World.* April 9. Accessed January 20, 2021. https://www.ren

ewableenergyworld.com/baseload/la-revolucion-energetica-cubas-energy-revolution/#gref.

Hausman, William, and John L. Neufeld. 2017. "The Cuban Electric Company, 1922–1960: From American Subsidiary to State Owned Enterprise." Draft paper for the European Business History Association Annual Meeting.

Juventud Rebelde. 2007. "Banca cubana puede asumir todos los créditos para el pago de los equipos electrodomésticos." December 4. Accessed January 20, 2021. http://www.juventudrebelde.cu/cuba/2007-12-04/banca-cubana-puede-asumir-todos-los-creditos-para-el-pago-de-los-equipos-electrodomesticos.

Käkönen, Mira, Hanna Kaisti, and Jyrki Luukkanen. 2014. *Energy Revolution in Cuba: Pioneering for the Future?* Turku: Finland Futures Research Centre, University of Turku.

MINCEX. n.d. *Cuba: Portfolio of Opportunities for Foreign Investment 2018–2019*. Ministerio del Comercio Exterior y la Inversión Extrajera.

MINCEX. 2017. *Cuba: Portfolio of Opportunities for Foreign Investment 2017–2018*. Ministerio del Comercio Exterior y la Inversión Extrajera. Accessed January 20, 2021. https://www.cubabusinessreport.com/portfolio2017-18.pdf.

Morales, Emilio. 2017. "Investment Report: Nickel, Oil, and Sustainable Energy." *Cubatrade*. March 1. Accessed January 20, 2021. www.cubatrademagazine.com/2637-2/.

Nachmany, Michal, Sam Fankhauser, Jana Davidová, Nick Kingsmill, Tucker Landesman, Hitomi Roppongi, Philip Schleifer, Joana Setzer, Amelia Sharman, C. Stolle Singleton, Jayaraj Sundaresan, and Terry Townshend. 2015. Climate Change Legislation in Cuba. Grantham Institute, London School of Economics.

New Energy Events. 2018. "Havana Energy Announces Second Biomass Plant in Cuba." February 28. Accessed January 20, 2021. http://newenergyevents.com/havana-energy-announces-second-biomass-plant-in-cuba/.

PCC. 2017. "Plan Nacional de Desarrollo Económico y Social hasta 2030: Propuesta de Visión de la Nación, ejes y sectores estratégicos." *Documentos del 7mo. Congreso del Partido aprobados por el III Pleno del Comité Central del PCC el 18 de mayo de 2017 y respaldados por la Asamblea Nacional del Poder Popular el 1 de junio de.*

Revista Eólica y de Vehículo Eléctrico. 2014. "Cuba impulsa las energías renovables en Isla de la Juventud." October 9. Accessed January 21, 2021. https://www.evwind.com/2014/10/09/isla-de-la-juventud-poligono-de-uso-de-energias-renovables/.

Rodríguez, José Luis. 2016. Interview with Helen Yaffe. December. Havana.

Sáenz, Tirso W. 2005. Interview with Helen Yaffe. January 7. Havana.

Seifried, Dieter. 2013. "Cuban Energy Revolution – A Model for Climate Protection?" Büro Ö-quadrat Report. Accessed January 12, 2021. http://www.oe2.de/fileadmin/user_upload/download/Energierevolution_Cuba_eng.pdf.

Suárez, José Antonio, Pedro Anibal Beatóna, Ronoldy Faxas Escalona, and Ofelia Pérez Montero. 2012. "Energy, Environment and Development in Cuba." *Renewable and Sustainable Energy Reviews* 16 (5): 2724–2731.

The Economist. 2017. "The Miracle of marabú, Cuba's Wonderful Weed." June 1. Accessed January 20, 2021. www.economist.com/news/americas/21722896-nuisance-can-be-used-light-barbecues-and-generate-electricity-miracle-marab.

US Department of Commerce. 1956. *Investment in Cuba: Basic Information for United States Businessmen.* Washington, DC: GPO.

Voss, Michael. 2016. "China, Cuba Cooperate in Developing Renewable Energy." *CGTN*. September 26. Accessed January 20, 2021. https://www.youtube.com/watch?v=ZbxECkx1J1k.

Yaffe, Helen. 2005. "Hurricane Dennis: When the Going Gets Tough . . ." *Fight Racism! Fight Imperialism!* 186 August / September. Accessed January 20, 2021. https://www.revolutionarycommunist.org/.

Chapter 13

The Foundation, Evolution, and Significance of Law as Part of Cuba's Adaptive Governance of Hazard Response

Shawn H. E. Harmon and Emily J. Kirk

Like the twenty-eight other small, island-based countries in the Caribbean, Cuba faces a seasonal onslaught of extreme weather (i.e., most typically hurricanes and tropical storms, with related flooding, drought, fire, and damage). Fidel Castro's revolutionary leadership faced its first significant natural hazard in 1963 when Hurricane Flora, only a Category 2 storm, caused some 1,200 deaths and enormous material loss. This event prompted important changes to the fledgling Popular Defense system, itself an evolution of the National Revolutionary Militia, which were formed to protect the people and the Revolution from internal and external acts of aggression. Further natural hazards followed, and the Popular Defense became the National Civil Defense (Defensa Civil, or DC), which itself has gradually shifted from its early emphasis on military threats to natural hazards. The DC has also evolved from a primarily response-oriented agency to a more holistic and highly effective hazard preparedness, mitigation, management, and recovery system (or collection of systems) with central control and clear lines of authority and responsibility.

With respect to increased threats, and to highlight the importance of this policy field for Cuba, from 1990 to 2014, Cuba suffered an average of two significant hazard impacts per year, resulting in an annual average of five deaths (Castellanos Abella and Wisner 2019; UNISDR 2015). While these losses are not inconsequential, they are almost trifling in comparison to those of Cuba's neighbors. For example, when Hurricane Matthew struck in 2016, Cuba experienced high winds and floods, evacuated 1.3 million people, and suffered some $2.5 billion in damage, but no deaths. By contrast, 40 deaths were reported in the United States and over 1,000 in Haiti (Kirk 2017). In the 2017 hurricane season, the region was lashed by multiple tropical storms and hurricanes. Cuba was hit hardest by Hurricane Irma, a Category 5 storm that strafed the island for three consecutive days, resulting in over $13 billion in damage, including damage to 158,554 homes, and forcing the evacuation of some 2 million people (out of a population of 11.3 million) (Zakrison et al. 2019). However, only ten deaths resulted (*Cubadebate* 2017; *Granma* 2017). This becomes more significant when contrasted to Puerto Rico's experience with Hurricane Maria, another Category 5 storm that hit in the same month, which resulted in some 2,975 deaths (George Washington University 2018). These very divergent experiences suggest that Cuba is doing something right and that it is useful for other countries and actors to be aware of what the island is doing and how it is doing it.

This chapter focuses on the legal elements of Cuba's hazard preparedness, mitigation, management, and recovery system(s). The legal framework—which particularizes, operationalizes, and fortifies the system(s) characteristics, and which has evolved and been refined over time—has been a critical component of Cuba's approach and a key factor in its success. As such, the framework is unpacked and assessed against notions of good governance. First, and importantly, the moral foundation of the legal framework is identified, which has endured, largely unchanged, since its inception. Second, the evolution of the framework (i.e., its key instruments and their shaping influence) is briefly described. Third, the legal framework from the perspective of good governance is assessed, focusing particularly on its reflection of elements of the ideal conception of "adaptive governance." While elements of the system and its development are undeniably unique to Cuba, much can be learned and much can be adapted for other jurisdictions, both comparable and incomparable to Cuba.

THE VALUE FOUNDATION OF THE CIVIL DEFENSE AND ITS LEGAL FRAMEWORK

To understand contemporary legal frameworks in Cuba, including the DC, one must appreciate the aims of the Revolution, and its ideological background based on the nationalist ideology of Cuban intellectual José Martí (1853–1895), which above all supported solidarity among Cubans. Following nearly four centuries of Spanish colonial rule and U.S. neo-colonialism beginning in 1898, extreme social-economic divisions between the wealthy (typically foreign) élite and the rest of the population were evident. The revolutionary movement, led by Fidel Castro, Raúl Castro, and Ernesto "Che" Guevara, began in the 1950s, gaining significant momentum in December 1956 with the landing of the *Granma* and subsequent rebellion. Notably, the revolutionary movement started in the east of the country—historically the poorest area—and spread westward. The rebellion, employing the revolutionary ideology of Martí, officially ended in January 1959, with the establishment of the revolutionary leadership, headed by Fidel Castro. The aims of the Revolution are encapsulated in the First Declaration of Havana, in September 1960, a polemic declaration that condemned imperialism and foreign interference with the democratic and developmental rights of Cubans, and which articulated a range of rights for the Cuban people (Castro 2007). Its overarching and explicit sense of solidarity was reaffirmed in the Second Declaration of Havana, in February 1962 (Ibid.), another polemic meant to rouse the population against foreign political interference and set the tone for the Revolution. This critical sense of solidarity was eventually enshrined in the Preamble and Articles 1 and 12 of the 1976 Constitution, as amended and highlighted by the island's only political party, the Cuban Communist Party (Partido Comunista de Cuba, or PCC) (Constitution of the Republic of Cuba 1976).

The 2019 Constitution articulates Cuba's contemporary political, social, legal, and economic position and objectives as they have evolved over the decades (Constitution of the Republic of Cuba 2019). It explicitly identifies liberty, equity, justice, equality, *solidarity*, and *wellbeing* as fundamental principles (Article 1). In keeping with these principles, it enumerates important rights, including rights to life, physical and moral integrity, justice, security, peace, health, education, culture, recreation, sport, and holistic development (Article 46), rights to free development of personality in keeping with respect, fraternity and solidarity (Article 47), and rights to a natural environment that is healthy and stable (Article 75).

Importantly, the 2019 Constitution also, contrary to many modern examples of Western constitutions and charters of rights, enumerates duties. Citizens have duties, *inter alia*, to respect others and not abuse individual rights; conserve, protect, and make rational use of resources; comply with environmental health standards and protect natural resources; and act in accordance with solidarity and respect for social coexistence (Article 90). The state also has duties; its objectives and duties include promoting sustainable development and social justice, guaranteeing the dignity of the people and safeguarding the natural, historical, and cultural patrimony of the nation (Article 13). Indeed, many of the individual rights contained in the 2019 Constitution are recognized as demanding state programs, such as social security, social housing, national healthcare, national primary-to-post-secondary education, and so on. (Articles 68–74). Notably, this can be contrasted with many Western constitutions, which tend to shy away from the direct imposition of obligations to provide (expensive) social programs, with many scholars viewing such provisions as aspirational and unenforceable (Constitution of the United States 2020; Dinan 2007).

It is evident in references to solidarity over time, as well as from its use throughout the 2019 Constitution, and from the collective substance of the rights and duties enumerated, that the Cuban notion of solidarity captures all three conceptions of solidarity identified in the literature (Scholz 2008), being:

- Social Solidarity: the bond, based on shared characteristics or conditions, between members of a community;
- Civic Solidarity: the bond, based on reciprocal obligations and a desire to protect the vulnerable, between members of a political community; and
- Political Solidarity: the bond, based on a shared moral relationship to a cause and a larger society, between members of an activist group or social movement.

All of them are characterized by a sense of people being bound together and having positive (moral) obligations to put others or the collective before themselves. They therefore support moral and social/policy imperatives to bear costs on behalf of others, and to craft policies and practices of distributive justice in pursuit of material equality (Ferdman and Kohn 2018). In other words, Cuban solidarity is a sense of connectivity that imposes obligations on all actors to *actively* or *positively* support others in various ways as the circumstances demand.

This communitarian socio-moral perspective has profoundly influenced the collective mentality of the Cuban people throughout their turbulent history, and it remains persistently relevant. It became an often-undeclared social foundation (or source of social cohesion and survival) when they were oppressed by the Spanish, and then by US neo-colonialism, including the US-supported dictator, Fulgencio Batista. It became more urgent and explicit as they were called upon to support the rebellion (1956–1958) and subsequent Revolution (1959–to date). And it became politically and legally ingrained as the revolutionary government reiterated it throughout its maturation and modern transformation (Aguirre 1984; Powell 2008; Seidman 2012). Its contemporary features, at least in relation to governmentality, have been described as a mixture of national pride, self-interest, and humanistic outreach, and its deployment as a calculated effort to overcome current structures of inequity and exploitation on a global scale by expanding new opportunities for marginalized communities (Huish 2014). Quite aside from precise definition, this enduring welfare-enhancing social norm has been harnessed, as will become apparent, by the legal framework supporting the DC. Indeed, the DC may be said to have operationalized it in a way that reaffirms, particularizes, and systematizes it in the hazard response setting.

Ultimately, the norm (solidarity) was a product of complex social forces, and the law shaping Cuba's hazard response architecture drew heavily on it to create a positive system of assessment, action, and responsibility in a complex setting characterized by a need to coordinate many actors and activities.[1] This marshaling of solidarity was and remains a central feature of the law. Its articulation is not only a primary objective of the law's "expressive function" (Mazzone 1999; Sunstein 1996), but it also strongly facilitates the framework's success by offering a socio-moral foundation for the imposition of hazard-related duties from top to bottom of the social strata.

THE CONSTITUENT INSTRUMENTS OF THE CIVIL DEFENSE'S LEGAL FRAMEWORK

While the social and policy significance of natural hazards is growing for countries all around the world in light of the consequences of climate change, they have an elevated significance for Cuba given its geographic, social, and economic circumstances. This significance imposes on Cuba (and others) the need for rational and systemic hazard responses. That Cuba has achieved the ability to cope with increasingly extreme and potentially devastating natural hazards, despite its modest resources and relatively fragile economic state, has prompted widespread interest (Gorry 2004; Thompson and Gaviria 2004). This interest has not often extended to the nature and content of the legal framework underpinning Cuba's widely lauded hazard preparedness, mitigation, management, and recovery system but that system *has* been shaped in no small part by law. Indeed, it was very early on entrenched in, and supported by, a strong legal framework that has been revisited and reformed in response to new knowledge and experience, and it is a legal framework that has been enforced assiduously (Defensa Civil 2017; Puig González et al. 2010a).

The first law to shape the DC system was the *National Defense Law 1966* (Puig González et al. 2010b), which states that the mission of the DC is, *inter alia*, to organize warning systems; develop evacuation plans and shelters to protect the population, particularly the vulnerable; and ensure the continuation of production and key services despite and during hazards. In keeping with the revolutionary emphasis on duties:

- all Ministries have a duty to account for hazard response in their financial and resource plans;
- all sub-national levels of government and all organizations have a duty to nominate DC coordinators and personnel with specific duties; and

- all organizations and individuals have a duty to fulfill all tasks assigned to them by the DC.

To achieve its mission and ensure compliance with stated duties, the law created the National Council of Civil Defense, chaired by the president and containing representatives from *all Ministries*. Thus, from the beginning, the scope of this field was set widely, with all areas of political/social action having a stake in its development and a role in its success. This scope, entrenched in law, demanded a much greater level of (systematized) coordination and cooperation than is seen in many other contemporary hazard response frameworks.

The *Civil Defense Organization Law 1976* (Puig González et al. 2010c) restructured the DC to reflect changes in the administrative structure of the nation (i.e., the number of provinces had risen from six in 1960 to 14, plus one special municipality—15 de facto provinces—and municipal governments across Cuba). This represents another unique feature of Cuba's hazard response system—it maps onto the entire public administration architecture, with elements and representatives at all levels of government. The president chairs the Council of Ministers and has command of the DC, acting through the Minister of Defense, and every ministry is represented on the council. Furthermore, every ministry must have Civil Defense personnel, and every provincial and municipal leader is also a Civil Defense Director (of their regional or local division as the case may be), with command of local assets. Ultimately, leadership is centralized at the national, provincial, and municipal levels, with clear responsibilities assigned to all actors and with lead roles assigned depending on the nature and scope of consequences of the hazard.

Given the uncertainty generated by the collapse of the USSR, Cuba next adopted the *National Defense Law 1994* (República de Cuba 1994). This articulates (Articles 3–5) a new defense doctrine—the "War of All the People" (*Guerra de Todo el Pueblo*)—and tasks the government with deploying Cuba's resources to ensure the uninterrupted defense of the nation (República de Cuba 1994). Though focused on the organization and functions of Defense Councils, the structure and use of military forces, and the general principles of preparing the country for security defense (Article 1), it also addresses civil defense, reaffirming the DC as a system of defensive measures (e.g., rescue, urgent repair, rehabilitation from contamination) carried out in peacetime and during "exceptional situations" (Article 9)[2] to protect the population and economy against destruction from enemies, natural disasters, other types of catastrophes, and environmental degradation (Article 111). Provinces, municipalities, and local defense zones are directed to have Defense Councils, which are prepared to act at all times in accordance with defense plans (Articles 28–30), including civil defense plans

(Articles 113–115). Specifically enumerated protective interventions include condition monitoring and the transmission of information, education of citizens on the rules of conduct to be followed,[3] distribution of means of protection against specific risks, protection of citizens in protective works, evacuation of people to safe or less threatened areas, actions to protect facilities, services, food supplies, and animals (Articles 11 and 116–117) (República de Cuba 1994). Importantly, the law stipulates in multiple provisions that compliance is compulsory and that actions and standards will be enforced.[4]

Thereafter, Cuba adopted the *System of Civil Defense Measures Law 1997* (Decreto Ley 170 1997), which affirms that a broad range of social organizations, institutions, and enterprises have responsibilities in relation to civil defense. It reiterates the DC hierarchical structure and clarifies the roles of the various actors, which currently include schools and universities,[5] local mitigation committees (which conduct risk mapping and simulation drills), and a range of hazard surveillance bodies.[6] It also, importantly, introduces a new mandate on all actors to identify and evaluate risks, hazards, and local vulnerabilities and to plan for their mitigation and/or management (Articles 2–7). In addition, it establishes the stages of hazard management (Article 13):

- Phase 1: Emergency (72 hours pre-event): When hazards such as hurricanes are predicted, the media begins broadcasting pre-packaged messages with up-to-date information and instructions, thereby putting the DC actors (e.g., Civil Defense Directors, workplace managers, physicians, etc.) on alert, who then review their plans and ensure that their respective assets are ready to mobilize.
- Phase 2: Alert (48 hours pre-event): In this phase, everyone mobilizes, with local, municipal, and provincial Civil Defense Centers becoming the locus for actions. Students are sent home and people await any evacuation orders directed at high-risk and vulnerable populations. Animals are moved and vulnerable crops are harvested.
- Phase 3: Alarm (event duration): Everyone must remain indoors at home or in shelters. The media continues to broadcast updated information, and designated actors work to ensure that core services remain functioning.
- Phase 4: Recovery (follow-on as-needed): Once the hazard has passed, leaders mobilize clean-up, repair, and provisions teams and begin certifying structures as safe for return. Local authorities calculate losses, and the president eventually deactivates the emergency system, and normal decision-making processes resume.

In furtherance of this law, multiple legal instruments were adopted. For example, *Decree 223* (Decreto No. 223 1997) stipulates that all resources in the country can be used in the interest of civil defense. The *Compatibility of Economic-Social Development and Civil Defense Law 1999* (Decreto Ley No. 262 1999) mandates that all state bodies and organizations, economic entities (e.g., manufacturers, importers, service-providers), and social institutions (e.g., research institutions) are expected to make their activities compatible with the interests of the DC (Article 4). In addition, as expressed in Article 5, they are "organs of consultation" in the defense of the nation, responsible to the National General Staff of the Civil Defense, and Civil Defense bodies at lower levels, and, as such, must assess risk and articulate measures for defense in relation to each investment or activity (Decreto Ley No. 262 1999). Generally, it is expected that investments will contribute to the preservation of sovereignty, defense, security, and internal order; incorporate defensive requirements for execution by the investor; raise the degree of protection in, *inter alia*, cases of catastrophes and natural disasters, promoting prevention through specified means[7]; and direct scientific and technological research at fulfilling the needs of civil defense (Article 2) (Decreto Ley No. 262 1999). Under the *Foreign Investment Law 2014* (which repealed the *Foreign Investment Law 1995*), foreign companies must follow all regulations relating to the protection against hazards, and each company's pre-investment study and investment plan must address hazard risk reduction and must be approved by DC headquarters (Ley No. 118 2014).

National Defense Council Directive 1/2005 further strengthens risk and harm reduction planning (República de Cuba, Consejo de Defensa Nacional 2005). Order 2 identifies the national disaster reduction process, which is characterized as a cohesive set of ideological, political, legal, economic-social, informational, communicative, internal, foreign relations, and military measures carried out to protect the population and economy during hazards that are foreseen by the DC system (República de Cuba, Consejo de Defensa Nacional 2005). Core elements of the national process are citizen engagement, multi-actor coordination, vulnerability identification, property protection, and maintenance of key social services throughout hazard events (Keyser and Smith 2009). It also categorizes a range of hazards to which actors *must* respond, offering a framework for assessment and demanding risk and harm reduction plans from various actors (Orders 1–2) (República de Cuba, Consejo de Defensa Nacional 2005). Moreover, disaster reduction planning must be integrated with socioeconomic planning and standards are set for planning processes and hazard reduction plans, including dissemination and communication expectations (Orders 8–10) (Ibid.). It also offers criteria for the activation of the different response phases (Castellanos Abella and Wisner 2019).

National Defense Council Directive 1/2007 establishes for the first time Risk Reduction Management Centers (RRMCs) in each province and municipality (Ibid.; Pardo Guerra et al. 2017). RRMCs are characterized as knowledge management spaces for the promotion of multidisciplinary risk and information management work and disaster prevention plan harmonization. A key ambition is to focus actor activities on reducing vulnerability to hazards and to generate local expertise and capacity. National Defense Council Directive 1/2010 describes the characteristics of the hazards that Cuba habitually faces (Ibid.). It also identifies DC Headquarters as responsible for communication with other levels of government and other entities, defines hazard-response time-frames, mandates the creation of disaster reduction plans by provinces, municipalities, local defense zones, and other economic entities and social institutions, and subordinates these entities and institutions to provincial and municipal defense councils in the hazard response and recovery phases. By Resolution 4/2017, the Chief of the National General Staff of Civil Defense emphasizes that disaster prevention and reduction is a governmental and non-governmental obligation (Ibid.). It appends a detailed methodological guide for the assessment of risk and vulnerability and the organization of harm reduction by state organs, national entities, and the local administrations. Key principles include:

- Feasibility: assessments and plans must respond to disaster reduction needs given the means available, and by using appropriate technology;
- Reproducibility: methods on which assessments rely must be clear and precise so studies can be reproduced;
- Monitorability: data and outcomes must be monitored over time for trends and progress, and a schedule for updating key datasets relied on must be included;
- Flexibility: assessments and plans must be sensitive to local needs and conditions;
- Comparability: despite the need for flexibility, assessments and plans must also be nationally comparable.

The Ministry of Science, Technology and Environment (Ministerio de Ciencias, Technología y Medio Ambiente, or CITMA), has the responsibility for coordinating these studies and ensuring that they are continually updated. This methodological guide has been described as a culmination of Cuba's hazard experience, a compilation of its disaster reduction framework instruments, and a comprehensive guide for organizing disaster reduction processes (Castellanos Abella and Wisner 2019).

An examination of the evolution and elements of Cuba's hazard preparedness, mitigation, management, and recovery system(s) reveals that the system has several key characteristics:

- stable core concepts, being solidarity, social justice, and human well-being;

- self-assessment and refinement to a current emphasis on forecasting, prevention, and proactivity;
- system support through widespread education and training and media deployment;
- clear, top-down but shared governance structures and processes;
- ongoing collaboration re: modeling and ever-improving forecasting;
- rational risk assessments, which include recognition of climate change (Alonso and Clark 2015); and
- comprehensive and resilient operational structures and practices.

Importantly, this system was shaped by, and reflected in, a strong legal framework that has been revised in response to new knowledge and experience. The law has:

- established clear lines of authority and articulated the mission of all actors;
- defined specific actor functions, authority, and flexibilities; and
- established mechanisms for inter-sectoral collaboration and institutional processes for learning.

Since the DC's beginnings in the early 1960s, it has, again, been guided by a strong sense of solidarity. As a result, through revisions to the legal framework, and with the help of the education, communication, and healthcare systems, a culture of risk awareness and collective safety has been encouraged, with a key political aim being to safeguard human life and wellbeing.[8] The result has been high levels of public trust (e.g., trust in the fact that DC systems and officials will prioritize health and wellbeing when fulfilling their tasks and will additionally take steps to preserve property when people are forced to leave property, often in vulnerable conditions) (Thompson and Gaviria 2004).

The significance of this legal framework and its active enforcement should not be underestimated when assessing the merit of the Cuban hazard response system (and its transportability). Laws can, and do, have a key role in shaping physical, social, and economic environments in ways that can promote or inhibit health, wellbeing, and equity, and they do so in many ways (e.g., they respond directly to inequitable and health-harming social conditions, alleviating or perpetuating them, or shaping influences in positive or disproportionate ways) (McGowan et al. 2019; Secretary's Advisory Committee for Healthy People 2030 2018). Furthermore, although it is possible to have actors work on collective plans, or to cooperate according to some established protocol, failures frequently occur, often when one of the actors does not fulfill its designated role, and often with cascading consequences (Conte and Dignum 2001). Cuba's hazard response system, enshrined in law, has done an admirable job of keeping many actors across government and across sectors on song and in tune. This is not a trivial achievement, and it begs the question: Why has it worked?

THE CIVIL DEFENSE'S LEGAL FRAMEWORK AS MODEL OF "ADAPTIVE GOVERNANCE"

In this section, we briefly demonstrate how the elements of "adaptive governance" (AG) are met by the Cuban hazard response system (and its legal framework).

At the outset, one must acknowledge the significance of the strong political will that has driven the evolution of the DC system. At base, the government has been willing to establish and invest in comprehensive and multi-level hazard preparedness, mitigation, management, and response plans in conjunction with efforts to improve infrastructure, including education and healthcare, invest in prevention and early warning systems, and develop effective means of communication, not only between critical institutions but across the polity (Kirk 2017; Thompson and Gaviria 2004). This level of commitment is not commonly paralleled in other jurisdictions. Indeed, the Cuban government has, to an unprecedented degree, been *willing* to (Kirk 2017):

- foreground human health, ecological sustainability, and human wellbeing in the face of increasingly frequent and ferocious natural hazards in all fields and levels of policymaking;[9]
- establish and invest in comprehensive and multi-level hazard preparedness, mitigation, management and response plans, linking same efforts to improve infrastructure;
- invest in early education, warning systems, health care, and the social determinants of health; and
- develop effective means of communication, not only between critical institutions but across the polity.

One *could* suggest that this level of sustained policy attention and investment is a consequence of Cuba's unique political structure and circumstances (e.g., a centralized and authoritarian government on a relatively small island nation), and that it cannot be reproduced in democratic states with quite different conditions. However, this suggestion unduly disregards the countervailing policy issues that *could have* distracted Cuba from this hazard-emphasizing path, but did not (i.e., the volatile and financially stressed times of the post-USSR Special Period, the need to cope with the United States' ongoing embargo and active efforts to pressure others into enforcing it more widely, etc.), and it unduly dismisses the failure of Western democratic states to engage effectively with hazard preparedness and the related policy issue of climate change.

However, neither of these phenomena have anything to do with the inherent character of political systems,[10] and as such should not be used to dismiss Cuba's success as unreproducible. On the contrary, we contend that Cuba's system may work because it more closely reflects a conception of AG, an arguably apolitical theoretical conception of good governance that is tailored for decision-making in contexts of uncertainty and complexity (Dietz et al. 2003), and which is widely viewed as being critical to successfully responding to developments in the ecological setting, and to pressures in the disaster risk reduction setting (Folke et al. 2005). In this respect, Cuba's success is not reliant on the nature of the political system from which it emerged, and it could be reproduced in other political contexts. At base, the aim of AG is to forge resilient governments, polities, and policies in the face of "wicked problems" such as adaptation to climate change conditions (Turini et al. 2010). We define resilience as the ability of a society to resist, absorb, accommodate, and recover from natural and human disturbances in a timely and efficient manner without undue infringement of social equity or loss of social character (Lizarralde et al. 2015; UNISDR 2009). In the institutional setting, it is defined as the ability to self-organize, learn, and adapt (Carpenter et al. 2001).

Importantly, AG is not a neutral institutional framework. In the social-ecological systems setting, AG seeks at least two value-salient normative outcomes, namely (1) to preserve the capacity of complex and dynamic ecosystems to generate services in support of human wellbeing and (2) to promote the desired social values and governance structures that enable sustainable ecological and social outcomes (Chaffin et al. 2014). While the literature on AG is extensive, evincing a variety of theoretical foundations and practical elements, governance approaches that contain the following characteristics can be understood as exemplifying AG:

- polycentricity and redundancy;
- complimentarity and integration;
- participatory and collaborativeness; and
- reflexivity and innovativeness.

We demonstrate below that these structural and operational characteristics are met by the Cuban hazard response system, as shaped by law, with the consequence that Cuban law can be said to have formed and enabled AG in Cuba.[11]

The first characteristic—polycentricity and redundancy—speaks to the need for the system to have multiple centers for decision-making and action and for key actors to therefore be "nested" in different levels of government and society.[12] When variously located, actors will have overlapping and complementary authority to respond to events, with responsibilities for taking decisions placed, ideally, on those as close to the problem (or locality of the hazard event) as possible. Broader concerns must also be capable of articulation and pursuit, so central authority is important, but power must be willingly shared. Ultimately, actors must constitute an interdependent system of relations with central and local actors having common goals formed as a result of multi-directional knowledge flows and trust, and an ability to defer authority to the other flexibly in real-time as circumstances warrant (Djalante et al. 2011).

The Cuban legal framework achieves this. For example, it recognizes and empowers actors at multiple levels and geographies (e.g., local agents, municipal leaders, provincial leaders, national ministers, the military hierarchy, non-governmental organizations, private or partnered commercial entities, etc.) (Defensa Civil 2017). In non-hazard circumstances, *all actors*, depending on their job and social placement, must contribute to institutional-level, local, and/or regional threat and vulnerability assessments and hazard reduction plans, and *all actors* must participate in hazard event preparation and training (Ibid.). When hazards arise, actors are expected to perform both common and bespoke duties, depending on their role and circumstances. Individuals not having formal DC positions have duties to safeguard family members and neighbors, to evacuate and follow orders, and to transmit evidence to appropriate authorities. Individuals with formal roles in the DC system have duties to mobilize resources, to safeguard the vulnerable, to secure property, and to report upward, and so on. Plans and leaders exist at the organizational level, the neighborhood level, the regional level, and so forth, and the power to make decisions rests at multiple levels in turn or simultaneously depending on the nature of the hazard and the phase of the hazard event. Ultimately, in keeping with characteristics noted by Lynch and Brunner (2010) and da Silveira and Richards (2013), multiple centers of power are created, and decisions must be exercised sensitive to conditions, and in pursuit of, a defined common interest. Operations, including communication, are managed centrally at the highest level,[13] but once surveillance systems identify the scope of the hazard, regional and local actors move to the fore, including directing the actions of military personnel. And because plans are co-designed, shared, and, to the extent possible, harmonized, actors can plug into operational activities relatively seamlessly. As such, it has been noted that the civil defense system "is everywhere" (Castellanos Abella and Wisner 2019), with responsibilities for hazard harm reduction infusing even normal (everyday) functions.

The second characteristic of AG is complementarity and integration, which refers broadly to the mutual recognition of legal authority across bodies and the need to therefore connect (Arnold 2014). It is obviously closely associated with the first characteristic, but it implicates and relies on the formation and enablement of networks of diverse actors coordinating to tackle a common problem. In this regard, boundary organizations, bridging organizations, and epistemic organizations are important. A boundary organization is a place for decision-makers to meet to create boundary arenas for all actors. The National Civil Defense Council, which encompasses the president and all national ministers, and the National General Staff of Civil Defense, together represent such an organization. Other important boundary bodies are the Physical Planning Institute (Insituto de Planificación Física, or IPF) and National Institute of Housing (Instituto Nacional de la Vivienda, or INV), and, at the regional level, Provincial Civil Defense Command and the Provincial Directorate of Physical Planning. A bridging organization is an entity or process that creates opportunities for different actors at different levels to interact and undertake joint assessments (Garmestani et al. 2008). In Cuba, the legislatively created and empowered RRMCs are bridging organizations, tasked with bringing experts together to advance knowledge, generate capacity, and formulate civil defense plans in a harmonized format. An epistemic organization is an institution or community of different actors who come together to influence or implement a policy, with management of the operation often shared (Carlsson and Berkes 2005). In Cuba, the education system generally, including universities, which undertake research in support of Cuba's hazard response policy development, represents an epistemic network. Similarly, *Meteoro* is an epistemic phenomenon; it pulls in all citizenry on an annual basis to both improve implementation of hazard response plans and to refine those plans. Management and assessment of *Meteoro* is diffuse, with information flowing upward toward the policy peak.

The third characteristic is about the participatory and collaborative elements of the system. Participation in the system and its governance can be formal or informal, and it can rely on different levels and locales of deliberation (Fung 2006). Collaboration requires events or ongoing processes where diverse actors can pool their knowledge and/or resources to solve a problem (Holley 2010). Combined, and if persistently exercised, participation and collaboration foster legitimacy and trust, strengthen individual and system capacities, and encourage resilience (Cosens et al. 2017; Pahl-Wostl 2009). It does this in part by contributing to social capital factors that enable social engagement and cooperation in relation to desired goals (Djalante et al. 2011).

In Cuba, the Federation of Cuban Women (Federación de Mujeres Cubanas, or FMC), the Workers Central Union of Cuba (Central de Trabajadores de Cuba, or CTC), and the Committees for the Defense of the Revolution (Comités de Defensa de la Revolución, or CDR) are important vehicles for participation and collaboration, and they are a means by which the state and the people are drawn closely together (Guzmán and Sánchez 2017). Neighborhood organizations and committees meet regularly to discuss local issues, facilitating the development of community bonds, which in turn improve social capacity to cooperate and respond to hazards. Community engagement through CDRs plays a role in harm prevention by identifying structures that are vulnerable, and those that can function as shelters. This knowledge is shared upward with the IPF, which, in consultation with other bodies, such as the Institute of Meteorology (Instituto de Meteorología, or INSMET), identifies high-risk areas and zones that are suitable for populations and industries to settle and evacuate to (Lizarralde et al. 2015). In short, and because Cuba's system is shaped in no small part by local professionals, participation need not include direct individual engagement with central authorities; it can be at non-governmental, institutional, and entirely local levels. Nonetheless, there are opportunities to engage more broadly in Cuba.

The final characteristic of AG is reflexivity and innovativeness, which necessitates the resources, capacity, and legal authority to respond to change (Cosens et al. 2017; Pahl-Wostl 2009). This reflexivity has multiple levels—it must be aimed at social learning and systemic learning. Social learning is the appreciation of one's role in the broader fabric of the shared objective and of the consequences of one's action to the broader public (Scholtz and Stiftel 2005). Systemic learning is about adjusting actions and decision-making processes in response to known and emerging variables, surprises, and uncertainty. Model AG systems should contain sites for both types of learning. A common hurdle to realizing this feature of AG is the technical, logistic, and financial challenges of crafting and systematizing true reflexive practices and institutions (Djalante et al. 2011; Gunderson 1999). An absence of reflexivity results in the absence of social memory of past hazards (and recovery) and so weakened capacities (Colten and Sumpter 2009).

Cuba has taken steps to enhance both social learning/memory and technical/systemic learning through investment in a complex infrastructure that systematizes social, technical, and logistic considerations in response to a range of hazard events. With respect to social learning, all citizens are reminded through multiple avenues, including the media, of their roles, and of the consequences (to themselves, their fellow citizens, the nation, and the DC system itself) of not fulfilling them. Through education, training, and mass messaging values are affirmed, social memory enhanced, and individual commitment and capacity advanced. With respect to systemic learning and innovation, *Meteoro*, in addition to being a training exercise and a test of that social learning, is a process for assessing both micro and macro practices and changing needs. Another process for learning is the ongoing revision of institutional, local, municipal, provincial, and national hazard reduction plans, which revision is meant to draw on experiences acquired in confronting and recovering from hazards as they occur

(Defensa Civil 2017).[14] In support of this, the RRMCs are sites of reflexivity and innovation; they are spaces of knowledge, which promote multidisciplinary work around risk estimation and management, information sharing, and the harmonization of all regional plans (Ibid.). Ultimately, Cuba's legal framework supports a system where experiences are accumulated so that changing circumstances can be responded and change can be managed (Folke et al. 2005).

It is clear from the above that the legal framework supporting Cuba's DC system has the characteristics of AG. It also, importantly, addresses justice concerns insofar as it encourages government decisions to be transparent and accessible, and it requires government actions to be appropriate, meaning science-based and consistent.

CONCLUSION

Cuban legislation in this field has undeniably performed an important shaping role. It has cast a broad net regarding who has responsibilities for taking actions and protecting wellbeing prior to and during hazard events; established clear lines of authority and specific actor functions and flexibilities, adding over time duties relating to effective foresight and harm prevention; and established mechanisms and institutions for local capacity development, inter-sectoral collaboration, and institutional learning and memory. In addition to being a good example of adaptive governance, Cuba's hazard response system (and its legal framework) can be understood as one component of a broader suite of ecological and developmental policies and laws aimed at achieving social justice in an environment characterized by risk, uncertainty, and ecological limitations, the latter of which are still woefully under-acknowledged by many Western states (Stricker 2010). Its system has been recognized as a model internationally (Bodenheimer 2017; Friedman-Rudovsky 2013; UN Office for Disaster Reduction 2004; WWF 2006). And while that precise model may not be transposable to other countries, reference to the AG framework to which Cuba's model so closely adheres could help other countries emulate Cuba's success and move closer to sustainability.

NOTES

1. In this regard, it has been noted that the content and effectiveness of law is influenced not only by collective action in response to a social problem that demands social institutions, but by the way law facilitates (or undermines) non-legal systems already relevant to the problem, such as norms (Posner 2000; Sunstein 1996).

2. "Exceptional situations" are defined in Article 9 of the 1994 *National Defense Law* as temporary national or regional conditions that threaten the nation, population

or economy, including military aggression, natural disasters, and other catastrophes or circumstances that, by their nature or proportion, affect internal order, security, or stability (República de Cuba 1994).

3. This provision must be understood in the context of a long-standing policy framework supporting population education and training. In 1986, *Meteoro* was introduced. This is a two-day national hazard preparation exercise involving the entire population, including children. Participation is mandatory. In addition, all children are exposed to disaster preparedness in their schools, where they participate in drills, learn first-aid and survival skills, learn to grow herbal medicines and find food, and are provided with pamphlets with information about emergencies. This education and systematic dissemination of information proceeds through secondary and into post-secondary education, and is accompanied by communication through state-run broadcast and newsprint media (Isayama and Ono 2015).

4. It has been observed that Cuba has impressive rates of compliance and mobilization (Sims and Vogelmann 2002).

5. By 2020, there were 60 universities in Cuba, all of them public (QS Top Universities: Cuba 2020).

6. By 2018, active surveillance systems in Cuba addressed meteorological, hydrological, seismological, epidemiological, and epizoological issues, as well as drought, fire, plant health, radiation, and oil spills (Castellanos Abella and Wisner 2019).

7. Obligations imposed on investors include monitoring conditions such as wind, rain, storm surges, floods, seismic activity, fires, toxic or radioactive leaks, and so on, transmitting notices, darkening lights, assisting with protective construction works, protecting social infrastructure and services, preserving facilities, equipment, machinery, raw material, reserves of all kinds and finished products.

8. Scholars often referred to this as a "culture of safety" (Sims and Vogelmann 2002; Thompson and Gaviria 2004).

9. This is in keeping with the Adelaide Statement on Health in All Policies (World Health Organization 2010). For more on global health justice and health-in-all-policies approaches, see Harmon (2015) and Harmon et al. (2015).

10. Having said that, we do acknowledge that privatization of public services and decentralized management of natural resources, both of which are being aggressively pursued by neoliberal elements within and without Western governments, have a track-record of failure (Acheson 2006; Rijke et al. 2012).

11. In this regard, it has been noted that, because the law is pervasive in government action, the criteria for AG will also serve as criteria for any inquiry into the role of law (Cosens et al. 2017).

12. Nesting means that lower-level decision-makers have representation in high-level structures, thereby facilitating the sharing of information, the consistency of implementation, and the building of trust (Bodin and Crona 2009).

13. For example, Council of Ministers Ordinance No. 279, 19 March 2007, identifies the group of entities tasked with obtaining, analyzing, evaluating, processing, and issuing hydrological and meteorological information for the execution of protective measures, and CITMA (1999) Resolution 106, designates the INSMET as the only

institution authorized to issue through the media, warnings and forecasts of meteorological hazards and related information for the protection of human life, material goods, and the economy.

14. It is stipulated that affected organizations and territories entities will update their Disaster Reduction Plan both annually and post-rehabilitation phase of a hazard, considering new risks and vulnerabilities discovered (Defensa Civil 2017).

BIBLIOGRAPHY

Acheson, James M. 2006. "Institutional Failure in Resource Management." *Annual Review of Anthropology* 35 (1): 117–134.

Aguirre, Benigno E. 1984. "The Conventionalization of Collective Behaviour in Cuba." *American Journal of Sociology* 90 (20): 541–566.

Alonso, Gisela, and Ismael Clark. 2015. "Cuba Confronts Climate Change." *MEDICC Review* 17 (2): 10–13.

Arnold, Craig Anthony. 2014. "Adaptive Water Law." *University of Kansas Law Review* 62 (4): 1043–1090.

Bodenheimer, Rebecca. 2017. "Cuba's Response to Hurricane Irma Puts 'America First-ism' to Shame." *Havana Times*. September 20. Accessed March 17, 2020. https://havanatimes.org/opinion/cubas-response-to-hurricane-irma-puts-america-first-ism-to-shame/.

Bodin, Örjan, and Beatrice Crona. 2009. "The Role of Social Networks in Natural Resource Governance: What Relational Patterns Make a Difference?" *Global Environmental Change* 19 (3): 366–374.

Carlsson, Lars and Fikret Berkes. 2005. "Co-Management: Concepts and Methodological Implications." *Journal of Environmental Management* 75 (1): 65–76.

Carpenter, Steve, Brian Walker, J. Marty Anderies, and Nick Abel. 2001. "From Metaphor to Measure: Resilience of What to What?" *Ecosystems* 4 (8): 765–781.

Castellanos Abella, Enrique A., and Benjamin Wisner. 2019. "Natural Hazards Governance in Cuba." *Oxford Research Encyclopedia of Natural Hazard Science*. Accessed February 15, 2020. https://www.researchgate.net/profile/Enrique_Castellanos2/publication/330635389_Natural_Hazards_Governance_in_Cuba/links/5c4b321fa6fdccd6b5c85750/Natural-Hazards-Governance-in-Cuba.pdf.

Castro, Fidel. 2007. *The First and Second Declarations of Havana: Manifestos of Revolutionary Struggle in the Americas Adopted by the Cuban People*, 3rd edition. New York: Pathfinder.

Chaffin, Brian C., Hannah Gosnell, and Barbara A. Cosens. 2014. "A Decade of Adaptive Governance Scholarship: Synthesis and Future Directions." *Ecology and Society* 19 (3): 56–69.

CITMA. 1999. *Resolution 106*. Ministry of Science, Technology and Environment. December 6.

Colten, Craig E., and Amy R. Sumpter. 2009. "Social Memory and Resilience in New Orleans." *Natural Hazards* 48 (3): 355–364.

Concepción, José Raúl, Sergio Alejandro Gómez, and Deny Extremera San Martín. 2018. "Tarea Vida a debate en la Asamblea Nacional: Cuba ante las realidades del cambio climático." *Cubadebate*. December 16. Accessed March 1, 2020. http://www.cubadebate.cu/noticias/2018/12/16/tarea-vida-a-debate-en-la-asamblea-nacional-cuba-ante-las-realidades-del-cambio-climatico/#.XkLquWhKiUk.

Constitution of the Republic of Cuba. 1976. *Cuba's Constitution of 1976 with Amendments through 2002*. Accessed February 15, 2020. https://www.constituteproject.org/constitution/Cuba_2002.pdf?lang=en.

Constitution of the Republic of Cuba. 2019. *Cuba's Constitution of 2019*. Accessed February 24, 2019. https://www.constituteproject.org/constitution/Cuba_2019.pdf?lang=en.

Constitution of the United States. 2020. *United States of America's Constitution of 1789 with Amendments though 1992*. Accessed February 15, 2020. https://www.constituteproject.org/constitution/United_States_of_America_1992.pdf?lang=en.

Conte, Rosaria, and Frank Dignum. 2001. "From Social Monitoring to Normative Influence." *Journal of Artificial Societies and Social Simulation* 4 (2): 1–7.

Cosens, Barbara A., Robin K. Craig, Shana Lee Hirsch, Craig Anthony Arnold, Melinda H. Benson, Daniel A. Decaro, Ahjond S. Garmestani, Hannah Gosnell, J. B. Ruhl, and Edella Schlager. 2017. "The Role of Law in Adaptive Governance." *Ecology and Society* 22 (1): 1–30.

Council of Ministers Ordinance No. 279. 2007. March 19.

Cubadebate. 2017. "Información del Consejo de Defensa Nacional." September 29. Accessed February 15, 2020. http://www.cubadebate.cu/noticias/2017/09/29/informacion-del-consejo-de-defensa-nacional/#.XkLsVWhKiUk.

da Silveira, André, and Keith S. Richards. 2013. "The Link between Polycentrism and Adaptive Capacity in River Basin Governance Systems: Insights from the River Rhine and the Zhujiang (Pearl River) Basin." *Annals of the Association of American Geographers* 103 (2): 319–329.

Decreto Ley 170. 1997. *Del Sistema de Medidas de Defensa Civi*. May 8. Accessed February 15, 2020. https://www.ifrc.org/docs/IDRL/Cuba%20Defensa%20Civilcub82158.pdf.

Decreto Ley No. 262. 1999. *Reglamento para la compatibilización del desarrollo económico-social del país con los intereses de la defensa*. May 14. Accessed February 15, 2020. https://www.onrm.minem.cu/registro/pdf/DL%20262.pdf.

Decreto No. 223. 1997. "De la reserva militar de medios y epuipos de la economianacional." In *Gaceta Oficial de la República de Cuba*, 506–508. Havana: Edición Ordinaria.

Defensa Civil. 2017. *La Defensa Civil Cubana*. Havana: Casa Editorial Verde Olivo.

Dietz, Thomas, Elinor Ostrom, and Paul C Stern. 2003. "The Struggle to Govern the Commons." *Science* 302: 1907–1912.

Dinan, John. 2007. "The Meaning of State Constitutional Education Clauses: Evidence from the Constitutional Convention Debates." *Albany Law Review* 70 (3): 927–981.

Djalante, Riyanti, Cameron Holley, and Frank Thomalla. 2011. "Adaptive Governance and Managing Resilience to Natural Hazards." *International Journal of Disaster Risk Science* 2 (4): 1–14.

Ferdman, Avigail, and Margaret Kohn. 2018. "The Theory and Politics of Solidarity and Public Goods." *Critical Review of International Social and Political Philosophy* 21 (5): 545–553.

Folke, Carl, Thomas Hahn, Per Olsson, and Jon Norberg. 2005. "Adaptive Governance of Social-Ecological Systems." *Annual Review of Environmental Resources* 30 (1): 441–473.

Friedman-Rudovsky, Jean. 2013. "Hurricane Tips from Cuba." *New York Times*. July 29. Accessed March 17, 2020. https://www.nytimes.com/2013/07/30/science/hurricane-tips-from-cuba.html.

Fung, Archon. 2006. "Varieties of Participation in Complex Governance." *Public Administration Review* 66: 66–75.

Garmestani, Ahjond S., Craig R. Allen, and Heriberto Cabezas. 2009. "Panarchy, Adaptive Management and Governance: Policy options for Building Resilience." *Nebraska Law Review* 87 (4): 1036–1054.

George Washington University. 2018. *Project Report: Ascertainment of the Estimated Excess Mortality from Hurricane María in Puerto Rico.* Accessed February 15, 2020. https://razoncienciaspr.files.wordpress.com/2018/08/informe-universidad-george-washington.pdf.

Gorry, Conner. 2004. "UN Lauds Cuba as Model of Hurricane Preparedness." *MEDICC Review*. Accessed February 15, 2020. http://www.medicc.org/publications/medicc_review/1004/pdfs/Top%20Story.pdf.

Granma. 2017. "Defense Councils mobilized to prepare for Hurricane Irma." September 6. Accessed February 15, 2020. http://en.granma.cu/cuba/2017-09-06/defense-councils-mobilized-to-prepare-for-hurricane-irma.

Gunderson, Lance. 1999. "Resilience, Flexibility and Adaptive Management—Antidotes for Spurious Certitude?" *Conservation Ecology* 3 (1): 7.

Guzmán, J. Sandoval, and D. Candabat Sánchez. 2017. "The Cuban System of Disaster Risk Reduction and its Applicability to Seismic Hazards." 16th World Conference on Earthquake Engineering. Accessed March 17, 2020. https://www.wcee.nicee.org/wcee/article/16WCEE/WCEE2017-3565.pdf.

Harmon, Shawn. 2015. "In Search of Global Health Justice: A Need to Reinvigorate Institutions and Make International Law." *Health Care Analysis* 23: 352–375.

Harmon, Shawn H.E., Lawrence Gostin, Liz Grant, John Gillies, and Graeme Laurie. 2015. "Imagining Global Health with Justice: Ebola, Impoverished People and Health Systems." *Medical Law International* 15 (1): 3–18.

Holley, Cameron. 2010. "Removing the Thorn from New Governance's Side: Examining the Emergence of Collaboration in Practice and the Roles for Law, Nested Institutions and Trust." *Environmental Law Reporter* 40 (7): 10656–10686.

Huish, Robert. 2014. "Why does Cuba 'Care' so much? Understanding the Epistemology of Solidarity in Global Health Outreach." *Public Health Ethics* 7 (3): 261–276.

Isayama, Kenji, and Naoya Ono. 2015. "Steps Towards Sustainable and Resilient Disaster Management in Japan: Lessons from Cuba." *International journal of Health System & Disaster Management* 3 (2): 54–60.

Keyser, Jonathan, and Wayne Smith. 2009. "Disaster Relief Management in Cuba: Why Cuba's Disaster Relief Model is Worth Careful Study." International Policy Report. Washington: Center for International Policy.

Kirk, Emily. 2017. "Dealing with the Perfect Storm: Cuban Disaster Management." *Studies in Political Economy* 98 (1): 93–103.

Ley No. 118. 2014. *Ley de la Inversión Extranjera*. March 29. Accessed February 15, 2020. http://www.onbc.cu/uploads/media/page/0001/01/eef3049ed87cb52a254b39719250268c55739dac.pdf.

Lizarralde, Gonzalo, Arturo Valladares, Andres Olivera, Lisa Bornstein, Kevin Gould, and Jennifer Duyne Barenstein. 2015. "A Systems Approach to Resilience in the Built Environment: The Case of Cuba." *Disasters* 39: s76–s95.

Lynch, Amanda H., and Ronald D. Brunner. 2010. "Learning from Climate Variability: Adaptive Governance and the Pacific ENSO Applications Center." *Weather, Climate, and Society* 2 (4): 311–319.

Mazzone, Jason. 1999. "When Courts Speak: Social Capital and Law's Expressive Function." *Syracuse Law Review* 49 (3): 1039–1066.

McGowan, Angela K., K. T. Kramer, Joel B. Teitelbaum, and Donna E. Levin. 2019. "Healthy People: The Role of Law and Policy in the Nation's Public Health Agenda." *The Journal of Law, Medicine & Ethics* 47: 63–67.

Pahl-Wostl, Claudia. 2009. "A Conceptual Framework for Analysing Adaptive Capacity and Multi-Level Learning Processes in Resource Governance Regimes." *Global Environmental Change* 19 (3): 354–365.

Pardo Guerra, Ramón, Luis A. Macareño Velíz, Albo Parra Salinas, Gloria Gely Martínez, Wilfredo Cobas Dávila, Raúl R. Coasta Gravalosa, Humberto de la Rosa, Noel Pérez Lozano, and Marbelis Rodríguez Azahares. 2017. *Guía metodológica para la organization de desastres*. Accessed March 1, 2020. https://www.preventionweb.net/files/59362_guiametodologicaparaorganizacionrrd.pdf.

Posner, Eric. 2000. *Law and Social Norms*. Cambridge: Harvard University Press.

Powell, Kathy. 2008. "Neoliberalism, the Special Period and Solidarity in Cuba." *Critique of Anthropology* 28 (2): 177–197.

Puig González, Miguel Ángel, José Ernesto Betancourt Lavastida and Rolando Álvarez Cedeño. 2010a. *Fortalezas frente a huracanes*. Havana: Editorial Científico.

Puig González, Miguel Ángel, José Ernesto Betancourt Lavastida and Rolando Álvarez Cedeño. 2010b. "Ley No. 1194. Sobre la organización de la Defensa Civil de la República de Cuba." In *Fortalezas frente a huracanes*, 153–156. Havana: Editorial Científico.

Puig González, Miguel Ángel, José Ernesto Betancourt Lavastida and Rolando Álvarez Cedeño. 2010c. "Ley No. 1316. Sobre el perfeccionamiento de la estructura organizativa de la Defensa Civil." In *Fortalezas frente a huracanes*, 157–162. Havana: Editorial Científico.

QS Top Universities: Cuba. 2020. Accessed February 17, 2020. https://www.topuniversities.com/where-to-study/latin-america/cuba/guide.

República de Cuba. 1994. *Ley No. 75 de la Defensa Nacional*. December 21. Accessed February 15, 2020. https://www.cubahora.cu/uploads/documento/2017/08/10/ley-no-75-de-la-defensa-nacional.pdf.

República de Cuba, Consejo de Defensa Nacional. 2005. *Directiva No. 1 Del Vicepresidente del Consejo de Defensa Nacional. Para la planificación y preparación del país par alas situaciones de desastres*. June 1. Accessed February 17, 2020. http://www.sld.cu/galerias/pdf/sitios/desastres/directiva_vp_cdn_sobre_desastres.ultima_version.pdf.

Rijke, Jeroen, Rebekah Brown, Chris Zevenbergen, Richard Ashley, Megan Farrelly, Peter Morison, and Sebastiaan van Herk. 2012. "Fit-for-Purpose Governance: A Framework to Make Adaptive Governance Operational." *Environmental Science & Policy* 22: 73–84.

Scholz, John T., and Bruce Stiftel, editors. 2005. *Adaptive Governance and Water Conflict: New Institutions for Collaborative Planning*. Washington: Resources for the Future Press.

Scholz, Sally J. 2008. *Political Solidarity*. University Park, PA: Pennsylvania State University Press.

Secretary's Advisory Committee for Healthy People 2030. 2018. *Issue Briefs to Inform Development and Implementation of Healthy People 2030: Submitted to the Secretary of the US Department of Health and Human Services*. Accessed January 21, 2020. https://www.healthypeople.gov/sites/default/files/HP2030_Committee-Combined-Issue%20Briefs_2019-508c.pdf.

Seidman, Sarah. 2012. "Tricontinental Routes of Solidarity: Stokely Carmichael in Cuba." *Jounral of Transnatinal American Studies* 4 (2): 1–25.

Sims, Holly, and Kevin Vogelmann. 2002. "Popular Mobilization and Disaster Management in Cuba." *Public Administration and Development* 22 (5): 389–400.

Stone, Richard. 2018. "Cuba embarks on a 100-year plan to protect itself from climate Change." *Science*. January 10. Accessed March 17, 2020. https://www.sciencemag.org/news/2018/01/cuba-embarks-100-year-plan-protect-itself-climate-change.

Stricker, Pamela. 2010. "Bringing Social Justice Back In: Cuba Revitalises Sustainable Development." *The International Journal of Justice and Sustainability* 15 (2): 185–197.

Sunstein, Cass R. 1996. "On the Expressive Function of Law." *University of Pennsylvania Law Review* 144 (5): 2021–2053.

Thompson, Martha, and Izaskun Gaviria. 2004. *Weathering the Storm: Lessons in Risk Reduction from Cuba*. Boston, MA: Oxfam America.

Turini, Alex, Danirla Cristofoli, Francesca Frosini, and Greta Nasi. 2010. "Networking Literature about Determinants of Network Effectiveness." *Public Administration* 88 (2): 528–550.

UN Office for Disaster Reduction. 2004. "Cuba: A Model in Hurricane Risk Management." September 14. Accessed October 21, 2020. https://www.un.org/press/en/2004/iha943.doc.htm.

UNISDR. 2009. *Terminology on Disaster Risk Reduction*. UN International Strategy for Disaster Reduction. Accessed March 17, 2020. https://www.unisdr.org/files/7817_UNISDRTerminologyEnglish.pdf.

UNISDR. 2015. "Annual Report 2015: 2014–2015 Biennium Work Programme Final Report. The United Nations Office of Disaster Risk Reduction." Accessed February 14, 2020. https://www.unisdr.org/files/48588_unisdrannualreport2015evs.pdf.

World Health Organization. 2010. *Adelaide Statement on Health in All Policies*. Accessed March 1, 2020. https://www.who.int/social_determinants/hiap_statement_who_sa_final.pdf.

WWF. 2006. *World Wildlife Fund: Living Planet Report 2006*. Accessed March 26, 2020. https://wwfint.awsassets.panda.org/downloads/lpr_living_planet_report_2006.pdf.

Zakrison, Tanya L., Davel Milian Valdés, and James Shultz. 2019. "The Medical, Public Health and Emergency Response to the Impact of 2017 Hurricane Irma in Cuba." *Disaster Medicine and Public Health Preparedness* 14 (1): 10–17.

Chapter 14

Timeline

Important Events in Cuban Disaster Management and Climate Change Adaptation

Jessica Hirtle

Since the beginning of the Revolution in 1959, Cuba's disaster preparedness and response measures have continued to develop and advance. While the frequency and intensity of natural disasters have only continued to increase, so too have Cuba's efforts to mitigate them. From the formalization of the Civil Defense to the implementation of Risk Reduction Management Centers (RRMCs), Cuba has steadily evolved to effectively prepare and protect its population, reduce hazards, mitigate damage, and minimize lives lost in times of disaster. The timeline below provides a chronological overview of the impacts of major meteorological events in Cuba, as well as the development of the island's disaster preparedness, response, and risk reduction strategies and significant events in climate change adaptation.

1959 The Military Organization of Industries (Organización Militar Industrial, or OMI) was created to protect economic and political targets (Castellanos Abella and Wisner 2019).

1960 Hurricane Donna, a Category 4 storm on the Saffir–Simpson scale, brushed northern Cuba and forced almost 8,000 people to evacuate inland to other houses or state-run shelters (Allison 1960; Puig González et al.).

1962 In February, the OMI was transformed into the Central Headquarters of Popular Defense. On July 31, Raúl Castro, the minister of the Revolutionary Armed Forces (Fuerzas Armadas Revolucionarias, or FAR), and the Coordination, Execution and Inspection Boards (Juntas de Coordinación, Ejecución e Inspección, or JUCEI) from provinces and municipalities across the country met to determine the responsibilities of the Civil Defense (Defensa Civil, or DC). This is widely considered to be the beginning of Cuba's Civil Defense system. In September, the first 1,000 people began their studies at the National School of Popular Defense (later known as the National School of Civil Defense) (Castellanos Abella and Wisner 2019; Puig González et al. 2010).

1963 In September and October, Hurricane Flora hit Cuba. A Category 2 storm, Hurricane Flora impacted 30,000 homes, 175,000 people were forced to evacuate, and 1,200 people were killed. The total cost of damage was between $300 and $500 million. After the hurricane, the Popular Defense's role shifted from protecting the population against military attack to responding to natural disasters (Bermejo 2006; Bohning 1964; Castellanos Abella and Wisner 2019; Kirk 2017; Puig González et al. 2010).

1964 Hurricanes Cleo (August), Hilda (September), and Isbell (October) struck Cuba and resulted in an economic loss of approximately $100 million. Hurricane Cleo affected fifty homes, thousands were evacuated, but no casualties were reported. Hilda was registered as a tropical storm when it crossed western Cuba. The storm caused heavy rains and flooding, forcing over 500 families to evacuate, but there were no fatalities. Hurricane Isbell also impacted western Cuba. Crops and livestock were heavily impacted, houses were damaged, thousands of people had to evacuate their homes, and three people were killed. There was an estimated loss of $20 million. In order to help the island recover from Flora and damaged inflicted from other hurricanes, the USSR donated a prefabricated housing factory (Bohning 1964; Puig González et al. 2010).

1965 The Institute of Meteorology (Instituto de Meteorología, or INSMET) was established to monitor and detect potential weather-related hazards (Castellanos Abella and Wisner 2019).

1966 In June, Hurricane Alma destroyed 13,400 acres of bananas and 18,090 acres of corn crops. Over 100,000 people were evacuated from danger, 78 were injured, and 12 people lost their lives in Havana, Las Villas, and Isla de Pinos (which was renamed Isla de la Juventud in 1978). In July, Law No. 1194 was passed to outline the mission and objectives of the Civil Defense, which included organizing warning systems, evacuation plans, transportation, and communication services, in the event of a disaster, and continuity planning for industries. In September and October, Category 3 Hurricane Inez hit Cuba. A total of 254,323 Cubans were evacuated in the provinces of Oriente, Camagüey, Las Villas, and Havana, 600 homes were affected, 40 people were injured, and four were killed (Castellanos Abella and Wisner 2019; Puig González et al. 2010; *Spokane Daily Chronicle* 1966; *The Fort Scott Tribune* 1966).

1967 Materials on disaster preparedness were introduced into the education system. The materials were first implemented into primary levels and then expanded into secondary, pre-university, and university levels (Castellanos Abella and Wisner 2019).

1968 Hurricane Abby impacted western Cuba in June, which resulted in the evacuation of 5,000 people in Guane. Four months later, Hurricane Gladys passed through western Cuba. In Pinar del Río, 10,288 individuals were evacuated from their homes and 32,293 animals were transported to higher ground. The hurricane resulted in minor damage, including some flooding and affected crops (Puig González et al. 2010).

1969 Hurricane Camille struck the province of Pinar del Río in August. The hurricane was registered as a Category 2 as it passed over Guane with winds surpassing 150 km/h. 32,000 people were evacuated in Pinar del Río province and Isla de Pinos, 20,000 animals were protected, and 217 houses were impacted (Puig González et al. 2010). In addition, material related to preparation for disasters was bolstered throughout the education system.

1970 In May, Hurricane Alma destroyed 117 tobacco houses and over 300 homes as well as killed seven people. In August, Hurricane Celia caused heavy rain in Guane, which resulted in the evacuation of some residents from low-lying areas. Five deaths in Cuba were attributed to the hurricane (National Hurricane Center 1970; Puig González et al. 2010).

1971 Tropical Storm Laura impacted western Cuba in November. More than 30,000 people were evacuated from the provinces of Pinar del Río, Havana, and Isla de Pinos. The storm caused heavy rainfall, which impacted 226 houses and resulted in the death of one person (Puig González et al. 2010).

1972 Tropical Storm Agnes impacted Cuba in June. 695 homes were damaged, 37,390 people and 58,000 animals were evacuated, 14 people were injured, 1 person went missing, and 5 people were killed. That same year, the United Nations hosted the Conference on the Human Environment, its first major event on environmental issues (Puig González et al. 2010; Funes Monzote 2019).

1975 At the First Congress of the Cuban Communist Party (Partido Comunista de Cuba, or PCC), Fidel Castro celebrated the advances made in the field of science, including the creation and development of INSMET. In addition, the Congress approved the "scientific policy thesis," which indicated the need to protect the environment (Funes Monzote 2019; PCC 1975).

1976 Law No. 1316 was introduced to re-organize the structure of the Civil Defense to align with Cuba's new political and administrative system. In addition, the island's new Constitution was ratified, which expressly supported the protection of the population during times of disasters, noting that the government was responsible for the protection of lives and property in case of natural disasters. A National Commission for the Protection of the Environment and the Rational Use of Natural Resources (Comisión Nacional de Protección del Medio Ambiente y los Recursos Naturales, or COMARNA) was also set up under a new State Committee for Science and Technology (Castellanos Abella and Wisner 2019; Funes Monzote 2019; República de Cuba 1976).

1977 A mandate was passed that requires all adults in Cuba to complete civil defense training (Kirk 2017).

1979 In September, Hurricane David was located 50 kilometers northeast of Punta Lucrecia, Holguín and was forecasted to impact Holguín, Las Tunas, and Camagüey. However, the hurricane changed its course and moved away from Cuba. Soon after, Hurricane Frederic forced 1,200 Cubans to evacuate and over 250 homes were impacted; however, no deaths were reported (CENAIS 2002; National Hurricane Center 1979; Puig González et al. 2010).

1981 Cuba began to broadcast nightly weather updates. The objective of the Cuban government was to spread important weather-related information to reduce the number of lives lost in the event of a disaster. Hurricane Katrina hit Cuba in November. It was reported that thousands were evacuated, and two people were killed in Camagüey province. Damages were caused by heavy flooding, which resulted in a substantial loss of crops and over 4,600 homes were impacted. Law 33 for the Protection of the Environment and Rational Use of Natural Resources was also ratified (Funes Monzote 2019; Keyser and Smith 2009; National Hurricane Center 1981a, 1981b).

1982 Hurricane Alberto brought heavy rain to western Cuba in June, which resulted in severe flooding. Seventeen people were injured, and twenty-three fatalities were recorded (Puig González et al. 2010).

1985 In November, Category 2 Hurricane Kate resulted in the evacuation of 473,400 people and two deaths. 60,000 homes were also impacted by the storm (Castellanos Abella and Wisner 2019; Puig González et al. 2010).

1986 Meteoro, Cuba's national preparedness and response exercise was established and has since been held annually, usually over a weekend in May (Castellanos Abella and Wisner 2019).

1991 The National Group on Climate Change was established by INSMET (Thompson 2007).

1992 The president of each Popular Council became formally responsible for civil defense in their communities (Defensa Civil 2017).

1994 Law No. 75 reinforced the definitions, principles, and role of the DC. Law No. 75 mandated that the National Defense Council could mobilize the population and evacuate citizens in the event of a natural disaster. The Cuban government also approved the United Nations Framework Convention on Climate Change. Meanwhile, Tropical Storm Gordon impacted eastern Cuba in November. More than 30,000 people and 68,780 livestock were evacuated, and two deaths were reported. Guantánamo province was the most impacted region, with damaged roads, houses, and crops. Moreover, the new Ministry of Science, Technology and Environment (Ministerio de Ciencias, Technología y Medio Ambiente, or CITMA) was created and tasked with devising the National Environmental Strategy (Castellanos Abella and Wisner 2019; Gaceta Oficial de la República de Cuba 1995; Houck 2000; Puig González et al. 2010; Thompson and Gaviria 2004; UN Department of Humanitarian Affairs 1994).

1995 In Havana, the Physical Planning Institute (Insituto de Planificación Física, or IPP) passed an agreement to safeguard older buildings and construct any new infrastructures above sea level with a view to reduce vulnerabilities (Thompson and Gaviria 2004).

1996 Category 2 Hurricane Lili resulted in the evacuation of 421,200 individuals, but no lives were lost in Cuba (Castellanos Abella and Wisner 2019; Thompson 2007).

1997 Decree Law No. 170 reaffirmed DC strategies within a variety of institutions, the establishment of DC measures to safeguard the population and economy, the introduction of phases (normal, informative, alert, alarm) in the event of hurricanes, and increased funding for DC measures. Decree 223 also stated that all resources in Cuba could be commandeered and used in the event of a natural disaster. In terms of environmental protection, Law 81 was also instituted, which established the legal framework for a National System of Protected Areas, resulting in an extensive network of nature reserves, national parks, wildlife refuges, and other protected areas on land and in the sea (Castellanos Abella and Wisner 2019; Whittle et al. 2002; Whittle and Rey Santos 2006).

1998 Hurricane Georges (September) was recorded as a Category 1 hurricane with winds registered at 120 km/h when it hit Cuba. The hurricane resulted in the evacuation of 818,800 people and 750,000 animals and resulted in six deaths. Hurricane Mitch (October) also caused the evacuation of 50,600, but no deaths were recorded (Castellanos Abella and Wisner 2019; Kirk 2017; Puig González et al. 2010).

1999 Category 1 Hurricane Irene resulted in the evacuation of over 220,000 people, 4 fatalities, damage to more than 28,000 homes, and caused $149 million in damage. Decree Law 262 was implemented and stated that the DC headquarters must cross-check all investments and research plans regarding disaster management with a view to improve risk deduction. In December, the General Assembly adopted the International Strategy for Disaster Reduction (ISDR) (Castellanos Abella and Wisner 2019; Puig González et al. 2010; Thompson 2007).

2001 Hurricane Michelle (October), a Category 4, resulted in the evacuation of 783,400 people and 741,000 livestock, and five people lost their lives. 5,700 power line poles were destroyed and 1,849 schools, 2,573 tobacco houses, and over 166,000 homes were affected. The total cost of damage was estimated at $1.866 billion (Castellanos Abella and Wisner 2019; Kirk 2017; Puig González et al. 2010).

2002 In September, Cuba was struck by Hurricanes Isidore and Lili, both Category 2 storms. During Hurricane Isidore, 307,000 people were evacuated, while 385,300 were evacuated in Hurricane Lili. Hurricanes Isidore and Lili caused $713 million in damage, 92,291 homes were impacted, and one person lost their life. In October, Tropical Depression 14 caused the evacuation of 70,000 people in Cuba; however, no fatalities were reported. In addition, the Council of Ministers established a Renewable Energy Front to devise strategies and policies to improve the sustainable use of renewable energies (Castellanos Abella and Wisner 2019; National Hurricane Center 2002; Puig González et al. 2010; Thompson 2007).

2004 In August, Category 3 Hurricane Charley required 224,449 people to evacuate, and four people died. The hurricane damaged 312 health centers, 798 education centers, and 87,000 homes. Category 5 Hurricane Iván impacted Cuba in September. 2,226,066 individuals were evacuated, but no lives were lost. The two hurricanes resulted in the loss of $2.145 billion. In response to Hurricanes Charley and Iván, three recommendations were implemented in Cuba: Councils of Defense were made available to all levels to coordinate response and recovery actions, the focus on response was shifted to mitigation by re-defining the "plan for measures in cases of catastrophes" to a "plan for disaster reduction," and disaster reduction strategies were implemented within socioeconomic plans (Castellanos Abella and Wisner 2019; CNN 2019; Llanes Guerra 2010; Puig González et al. 2010).

2005 At the start of 2005, Cuba was impacted by the most severe drought in 100 years. In reaction to the World Conference on Disaster Reduction, the Hyogo Framework for Action was adopted in January. Directive 1/2005 was implemented by the National Defense Council in June to establish the use of disaster reduction plans and expand the application of risk assessment. Risk Reduction Management Centers (RRMCs) were also founded by the Civil Defense. The first Center was established in Guamá, which is located in the province of Santiago de Cuba. Hurricanes Dennis (July), Rita (September), and Wilma (October) caused an economic loss of $3.036 billion. Hurricane Dennis impacted 10 Cuban provinces, 175,615 homes were affected, and 16 fatalities resulted. In August, 8,000 people forced to evacuate in Pinar del Río due to heavy rain caused by Hurricane Katrina

(AFP 2005; Castellanos Abella and Wisner 2019; Democracy Now 2017; Llanes Guerra 2010; Puig González et al. 2010).

2006 In September, Hurricane Ernesto resulted in an economic loss of $95.1 million. It was also the "Year of the Energy Revolution" in Cuba, as the state initiated a major initiative to improve energy sovereignty and efficacy (Llanes Guerra 2010).

2007 The National Defense Council approved Directive 1/2007 to implement RRMCs in all provinces and municipalities in Cuba. Since 2007, Cuba's budget has allotted funding for disaster reduction activities. In October and November, Tropical Storm Noel caused heavy rain and forced 28,120 people to evacuate. 23,786 kilometers of roads and 59,826 houses were impacted, and total damages were estimated at $1.155 billion (Castellanos Abella and Wisner 2019; Puig González et al. 2010).

2008 Hurricanes Fay (August), Gustav (August), Ike (September), and Paloma (November) caused $9.7 billion of damage and 647,111 homes were damaged. Hurricane Gustav, a Category 4 storm, damaged 120,509 houses, 450,000 individuals were evacuated, and 60 percent of the population lost electricity. Hurricane Ike, also a Category 4 storm, caused damage to 511,259 homes and approximately 2.6 million people were evacuated. Hurricanes Gustav and Ike caused a total of 7 deaths. The legal system continued to increase and fortify legislation focused on disaster risk reduction; by 2008, these included 3 Laws, 7 Decree Laws, 13 Decrees, 21 Resolutions, and 1 Directive (Castellanos Abella and Wisner 2019; *CNN* 2019; Kirk 2017; Llanes Guerra 2008; Sánchez Miranda and Choonara 2010; *The Guardian* 2008).

2010 Directive 1/2010 was passed by the National Defense Council and solidified leadership roles within municipal and provincial defense councils, strengthened communication between the government and other entities, and defined the timing of disaster response in specific situations (Castellanos Abella and Wisner 2019).

2012 A methodological guide was published by the National Civil Defense headquarters, which organized disaster reduction-related processes and measures. In October, Hurricane Sandy caused significant damage in eastern Cuba. The Category 2 storm killed 11 people, 343,230 people were evacuated, more than 170,000 homes were damaged, and there was a total economic loss of $4.7 billion (Castellanos Abella and Wisner 2019; Grogg 2012; Rainsford 2012).

2015 Cuba supported and joined the Sendai Framework, developed by the UN Office for Disaster Risk Reduction, which commits to: "The substantial reduction of disaster risk and losses in lives, livelihoods and health and in the economic, physical, social, cultural and environmental assets of persons, businesses, communities and countries." Additionally, CITMA was commissioned to develop a State Plan on the issue of climate change, converting earlier scientific results into policy (UNISDR 2015).

2014 The introduction of Law No. 118 affirmed that foreign companies must follow rules to protect against natural hazards and disasters, pre-investment plans must include "disaster studies," and investment plans must include National Civil Defense-approved provisions for disaster risk reduction (Castellanos Abella and Wisner 2019).

2016 Hurricane Matthew, a Category 4 storm, struck Cuba in October. 316,000 people were evacuated, 46,706 houses were damaged, and $2.4 billion in damage resulted. However, no fatalities were reported (AFP 2016; Castellanos Abella and Wisner 2019).

2017 The National Civil Defense issued a guide to outline disaster reduction measures and strategies to evaluate risks and vulnerabilities. CITMA began to coordinate Hazard, Vulnerability and Risk (Peligro, Vulnerabilidad y Riesgo, or PVR) studies in every municipality in Cuba. *Tarea Vida* (Project Life) was also founded to confront the impacts of climate change and improve adaptation strategies. In September, Category 5 Hurricane Irma caused $13 billion in damage, impacted 158,554 homes, and caused a reported 10 deaths. More than 2 million people were evacuated before the storm made landfall (Castellanos Abella and Wisner 2019; Kilkenny 2017; Rodríguez Milán 2018; United Nations 2017).

2019 The Cuban Constitution was amended to include further commitment of the Cuban government and population to combat climate change. Article 16 "promotes the protection and conservation of the environment and the confrontation of climate change, which threatens the survival of the human species." In addition, Cuba's National Assembly passed the country's first Fisheries Law (*Ley de Pesca*) replacing the previous Decree Law 164 of 1996. The updated legislation centers on sustainable fisheries management with a view to improve food sovereignty, security, and conservation (Constitution of the Republic of Cuba 2019; Ley No.129/2019 Ley de Pesca 2020).

2020 In April, the government announced a national strategy for food sovereignty, and nutritional education, which was aligned with the national climate change strategy, *Tarea Vida*.

Beginning in the 1960s, revolutionary Cuba has expanded and streamlined its disaster preparedness and response strategies. Even though the acceleration of climate change and natural disasters has persisted, it is evident that Cuba has evolved to adapt and respond to, as well as mitigate, these challenges. The country has continued to prioritize prevention and risk reduction through the standardization and systematization of its disaster preparedness and response methods. These methods have culminated in Cuba's ability to prepare and react in times of disasters and effectively reduce casualties. Like the revolutionary project itself, Cuba's efforts in disaster management and confronting climate change have continued to evolve, meeting the contemporary demands with innovation and the resources available.

BIBLIOGRAPHY

AFP. 2005. "Flooding and Power Outages as 'Katrina' Batters Western Cuba." *TerraDaily*. Accessed January 28, 2021. https://www.terradaily.com/2005/050828175451.y7y367k2.html.

AFP. 2016. "Cuba: 316.000 Personnes Evacuées Avant L'Arrivé de L'Ouragan Matthew." *DH Les Sports*. Accessed January 28, 2021. https://www.dhnet.be/actu/monde/cuba-316-000-personnes-evacuees-avant-l-arrivee-de-l-ouragan-matthew-57f3389dcd70871fc4251e2d.

Allison, Grenville, Jr. 1960. "Slight Wind Shift Heightens Danger." *The Times-News*. September 1. Accessed January 28, 2021. https://news.google.com/newspapers?id=B_QZAAAAIBAJ&pg=6119,433570&dq=hurricane+donna&hl=en.

Bermejo, Pedro Mas. 2006. "Preparation and Response in Case of Natural Disasters: Cuban Programs and Experience." *Journal of Public Health Policy* 27(1): 13–21.

Bohning, Dan. 1964. "4 Storms in a Year Strain Cuban Economy." *The Miami Herald*. October 18. Accessed January 28, 2021. https://www.nhc.noaa.gov/archive/storm_wallets/atlantic/atl1964/isbell/news/mh1018.gif.

Castellanos Abella, Enrique A., and Benjamin Wisner. 2019. "Natural Hazards Governance in Cuba." *Oxford Research Encyclopedia of Natural Hazard Science* 1–39.

CENAIS. 2002. "Evaluating Risks from Natural Hazards: Cuba." National Center for Seismological Research, Havana Division. Accessed January 28, 2021. https://web.archive.org/web/20150402115154/http://tcdc2.undp.org/GSSDAcademy/SIE/Docs/Vol12/2Cuba.pdf.

CNN. 2019 "Hurricane Statistics Fast Facts." May 31. Accessed January 17, 2021. https://www.cnn.com/2013/05/31/world/americas/hurricane-statistics-fast-facts/index.html.

Constitution of the Republic of Cuba. 2019. *Cuba's Constitution of 2019*. Accessed February 24, 2019. https://www.constituteproject.org/constitution/Cuba_2019.pdf?lang=en.

Defensa Civil. 2017. *La Defensa Civil*. Havana: Casa Editorial Verde Olivo.

Democracy Now. 2017. "After Irma, A Look at Why Cubans are 15 Times Less Likely to Die from Hurricanes Than Americans." September 12. Accessed January 20, 2021. https://www.democracynow.org/2017/9/12/after_irma_a_look_at_why.

Funes Monzote, Reinaldo. 2019. "*Geotransformación*: Geography and Revolution in Cuba from the 1950s to the 1960s." In *The Revolution From Within: Cuba, 1959–1980*, edited by Michael J. Bustamente and Jennifer L. Lambe, 124–128. Durham, NC: Duke University Press.

Gaceta Oficial de la República de Cuba: Ministerio de Justicia. 1995. *Ley Número 75 de la Defensa Nacional*. Havana.

Grogg, Patricia. 2012. "Mental Health, Another Victim of Climate Change." *Inter Press Service News Agency*. November 23. Accessed January 15, 2021. http://www.ipsnews.net/2012/11/mental-health-another-victim-of-climate-change/.

Houck, Oliver A. 2000. "Environmental Law in Cuba." *Journal of Land Use and Environmental Law* 16 (1): 1–81.

Keyser, Jonathan, and Wayne Smith. 2009. "Disaster Relief Management in Cuba: Why Cuba's Disaster Relief Model is Worth Careful Study." International Policy Report. Washington: Center for International Policy.

Kilkenny, Carmel. 2017. "Cuba Recovering as Hurricane Maria Approaches." *Radio Canada International*. September 18. Accessed January 20, 2021. https://www.rcinet.ca/en/2017/09/18/cuba-recovering-as-hurricane-maria-approaches/.

Kirk, Emily. 2017. "Alternatives–Dealing with the Perfect Storm: Cuban Disaster Management." *Studies in Political Economy* 98 (1): 1–11.

Ley No. 129/2019 Ley de Pesca. 2020. *Gaceta Oficial de la República de Cuba*. February. Accessed April 15, 2020. https://www.gacetaoficial.gob.cu/es/decreto-1-de-2019-de-consejo-de-ministros.

Llanes Guerra, José. 2008. *Cuba: paradigma en la reducción de riesgo de desastres*. Accessed September 20, 2018. https://www.preventionweb.net/files/2558_CubaParadigmaenlareducciondeRiesgodedesastres.pdf.

Llanes Guerra, José. 2010. *Cuba Risk Reduction Management Centres: Best Practices in Risk Reduction*. Colombia: UNDP.

National Hurricane Center. 1970. "Preliminary Report Celia." Accessed January 20, 2021. https://www.nhc.noaa.gov/archive/storm_wallets/atlantic/atl1970-prelim/celia/prelim02.gif.

National Hurricane Center. 1979. "Preliminary Report: Hurricane Frederic." Accessed January 20, 2021. https://www.nhc.noaa.gov/archive/storm_wallets/atlantic/atl1979-prelim/frederic/prelim02.gif.

National Hurricane Center. 1981a. "Hurricane Katrina Passes over Cuba." Accessed January 20, 2021.https://www.nhc.noaa.gov/archive/storm_wallets/cdmp/dvd0141-jpg/1981/atlantic/katrina/postevent/report_havana1109.jpg.

National Hurricane Center. 1981b. "Preliminary Report: Hurricane Katrina." Accessed January 20, 2021. https://www.nhc.noaa.gov/archive/storm_wallets/atlantic/atl1981-prelim/katrina/prelim01.gif.

National Hurricane Center. 2002. "Tropical Depression Fourteen." Accessed January 20, 2021. https://www.nhc.noaa.gov/archive/2002/FOURTEEN.shtml.

PCC. 1975. *Informe del Comité Central del PCC al Primer Congreso*. Havana: Departamento de Orientación Revolucionaria del Comité Central del Partido Comunista de Cuba.

Puig González, Miguel Ángel, José Ernesto Betancourt Lavastida, and Rolando Álvarez Cedeño. 2010. *Fortalezas Frente a Huracanes (1959–2008)*. Havana: Editorial Científico-Técnica.

Rainsford, Sarah. 2012. "Hurricane Sandy: Cuba Struggles to Help Those Hit." *BBC News*. November 12. Accessed January 17, 2021. https://www.bbc.com/news/world-latin-america-20294006.

República de Cuba. 1976. *Constitución Política de 1976*. Accessed September 20, 2018. http://pdba.georgetown.edu/Constitutions/Cuba/cuba1976.html.

Rodríguez Milán, Yisell, and Danae González Del Toro. 2018. "Climate Change Brings Transformations in Cuba." *Granma*. April 12. Accessed January 15, 2021. http://en.granma.cu/cuba/2018-04-12/climate-change-brings-transformations-in-cuba.

Sánchez Miranda, Deybis, and Imti Choonara. 2010. "Hurricanes and Child Health: Lessons from Cuba." *Archives of Disease in Childhood* 96 (4): 328–329.

Spokane Daily Chronicle. 1966. "Alma Rakes Florida Coast: Thousands Are Evacuated." June 9. Accessed January 28, 2021. https://news.google.com/newspapers?id=bWVYAAAAIBAJ&sjid=rfcDAAAAIBAJ&pg=4048,2627926&dq=hurricane+alma&hl=en.

The Fort Scott Tribune. 1996. "Cuba Hit Hard by Hurricane." June 9. Accessed January 28, 2021. https://news.google.com/newspapers?id=B9YfAAAAIBAJ&sjid=J9kEAAAAIBAJ&pg=6082,2027066&dq=hurricane+alma&hl=en.

The Guardian. 2008 "Hurricane Gustav Hits Cuba." August 30. Accessed January 17, 2021. https://www.theguardian.com/world/2008/aug/30/naturaldisasters.caymanislands.

Thompson, Martha. 2007. "Lessons in Risk Reduction from Cuba." *Enhancing Urban Safety and Security: Global Report on Human Settlements*, 319. Accessed January 17, 2021. https://www.un.org/ruleoflaw/files/urbansafetyandsecurity.pdf

Thompson, Martha, and Izaskun Gaviria. 2004. *Cuba, Weathering the Storm: Lessons in Risk Reduction from Cuba*. Boston, MA: Oxfam America.

UN Department of Humanitarian Affairs. 1994. "Cuba-Tropical Storm Nov 1994 UN DHA Situation Reports 1-5." Accessed January 29, 2021. https://reliefweb.int/report/cuba/cuba-tropical-storm-nov-1994-un-dha-situation-reports-1-5.

UNISDR. 2015. "Sendai Framework for Disaster Risk Reduction." United Nations Office for Disaster Risk Reduction. Accessed September 20, 2019. https://www.unisdr.org/we/coordinate/sendai-framework.

United Nations. 2017. "Cuba Hurricane Irma: Three-Month Report." Accessed January 15, 2021. https://reliefweb.int/sites/reliefweb.int/files/resources/Three_month_report_Irma.pdf.

Whittle, Daniel J., Kenton C. Lindeman, and James T.B. Tripp. 2002. "International tourism and protection of Cuba's coastal and marine environments." *Tulane Environmental Law Journal* 16: 533–589.

Whittle, Daniel, and Orlando Rey Santos. 2006. "Protecting Cuba's Environment: Efforts to Design and Implement Effective Environmental Laws and Policies in Cuba." *Cuban Studies* 37: 73–103.

Conclusion

Over the last six decades of the Revolution, Cuba has transformed from a passive victim of natural disasters into an international paradigm of best practice in sustainable, inclusive, and resilient disaster prepared management. Its hard-won expertise and experience are now being applied and developed to help combat the new global challenge of climate change. This book has offered a broad overview of such disaster preparedness infrastructure and practices in Cuba. It has shown that what initially emerged as an empirical response to the devastation of early natural disasters quickly became an integral focus and approach that underpinned many of the Revolution's subsequent developments.

This volume has brought together scholars and policy makers, with diverse expertise in order to present a comprehensive overview of existing research into the now inextricably linked, and increasingly broad, fields of climate change and disaster preparedness and management in Cuba. The sub-topics explored in the preceding chapters all align with one or more of the following five interrelated pillars that serve to buttress Cuba's approach to disaster preparedness and management: preparation, innovation, resilience, inclusion, and education. This collection demonstrates that Cuba's commitment to uphold these five overlapping tenets as part of its natural disaster and, more recently, climate change strategy has resulted in an ongoing process of adaptation, evolution, and responsiveness that is both distinct on the global stage and highly effective.

The studies included here reveal that, of these tenets, the education of the wider population has played a particularly pivotal role in informing many of Cuba's responses to natural disasters and the need to manage their potentially devastating effects accordingly. As Emily J. Kirk discusses in the opening chapter, a commitment to constant improvement, information dissemination, large-scale mobilization, and mutual help—all of which have the preservation of life and the minimization of damage at their very core—are just aspects of what sets Cuba apart from its peers. At the very basis of this education-centered approach is the drive to communicate the risks posed by natural disasters and the provision of basic education in how to respond as an enduring means to empower the general population. The primacy of education is also reflected at the more specialist levels; several authors in this collection have highlighted the consistent encouragement of research-driven expertise and innovation. Indeed, long-standing research into natural hazards, natural disasters, and the challenges posed by climate change has historically allowed for Cuba to participate on a more equal footing with other, much larger countries in the realm of disaster preparedness.

Education also directly translates into better and more consistent levels of preparation. This idea is borne out by Guillermo Mesa Ridel's discussion on postgraduate training in disaster preparedness, which underlines the growing need for significant capabilities in knowledge management, as well as specialist knowledge itself. Mesa Ridel outlines the increasing integration of interdisciplinary education in disaster management with improved access, and clearer communication of knowledge, to the wider population and a systematic, easily replicable (and therefore exportable) approach. He also underscores the responsiveness of the national educational efforts with regard to the ever-shifting needs of the population and the problems created by climate change. Other chapters similarly shine a light on the pledge to embed long-term, considered, and meaningful preparation across as many disciplines and fields of study as possible. Most recently, this pledge has been codified at the highest levels and now underwrites all future-focused activities and development in Cuba.

Regarding innovation, this book's authors have shown how preparedness is constantly being improved by the ongoing search to find new ways of exploiting the country's existing resources. In innovating, resilience is also increased significantly. Resilience has gained added importance as climate change becomes the overriding threat with the potential to change Cuba's landscape, its population's livelihoods, and the natural environment to a dramatic extent. In discussing the paths to greater resilience, authors like Margarita Fernández et al. have highlighted the need for the greater integration of scientific and traditional knowledge bases to help recognize and respond to Cuba's specific socioeconomic, historical, and environmental conditions. It is argued that an understanding and (re)valorization of local traditions and their specific needs is particularly integral to producing sustainable systems that have a high degree of resilience. The move away from fossil fuels to more sustainable, less impactful energy sources is implicit in this focus on local specificities and greater resilience.

Ultimately, all of the aforementioned pillars feed into, and are permanently underpinned by, the quest for the maximum inclusion of the Cuban people and their enduring sense of duty to defend themselves against all threats. Chapters such as Hal P. Klepak's on Cuba's Armed Forces and Civil Defense system(s) have drawn attention to the overarching objective of the Revolution's early programs to involve a wide breadth of stakeholders. Subsequent drives for greater inclusion have translated into the way that mass mobilization activities are framed, such as *Bastión* and *Meteoro*; these drives have been clearly and carefully linked to Cuba's long history of struggle for independence in a way that emphasizes popular participation and collective action. This emphasis is perhaps unsurprising, since, as a number of authors have stressed, the centrality of inclusion also reflects the well-established culture of political, social, and cultural participation in Cuba. Furthermore, this volume has shed light on the fact that the clear, accessible, and repeatable communication of high-level complex research and knowledge to the general population is an increasingly fundamental tenet of the Revolution's commitment to inclusion.

As many of the authors have pointed out in their chapters, Cuba is very much at the forefront of experiencing the adverse and dangerous effects of climate change. Yet the country's long history of disaster preparedness and management has enabled it to develop myriad effective responses that are rooted at the community level but are protected and encouraged by the state. The continued focus on research geared toward a better understanding of the complexities of climate and disaster mitigation has been coupled with a finessing of approaches that were initially born out of necessity but which have yielded impressive results. The success of Cuba's approaches can largely be explained by its dedication to employ accessible, local resources in such a way as to protect the country's hard-won sovereignty.

Choosing to forge its own path in terms of responding to climate change and natural disasters more broadly has not meant, however, that Cuba has shied away from international collaboration in these respects. Far from it. As this volume makes clear, Cuba's strategies have consistently been developed with a view to sharing knowledge, expertise, and experience beyond the island's shores. The country's long-standing and impactful participation in disaster relief efforts across the globe offers proof that its home-grown systems and approaches constitute invaluable models that other nations would do well to emulate. With climate change now widely considered a global emergency, the need for such transnational cooperation and solidarity that transcends political differences becomes ever more pressing.

Index

About the Contributors

Emily J. Kirk is a research fellow in the Department of International Development Studies and an adjunct professor in the Faculty of Graduate Studies at Dalhousie University. She is the author of *Cuba's Gay Revolution: Normalizing Sexual Diversity through a Health-Based Approach* and co-editor of *Cuba's Forgotten Decade: How the 1970s Shaped the Revolution*. Her current research focuses on disaster preparedness and management in the global south.

Isabel Story is a senior lecturer in visual communications at Nottingham Trent University. Her research focuses on Cuban–Soviet relations and Cuba's position within the wider field of international socialism. Her previous publications include *The Soviet Influence on Cuban Culture*. She was also a co-editor for the volume, *Cuba's Forgotten Decade: How the 1970s Shaped the Revolution*.

Anna Clayfield is a senior lecturer in Spanish and Latin American studies at the University of Chester, the United Kingdom. She is the author of *The Guerrilla Legacy of the Cuban Revolution* (2019), which explores the perpetuation of a guerrilla ethos in Cuba since 1959, and is co-editor of *Cuba's Forgotten Decade: How the 1970s Shaped the Revolution* (2018). Her research centers on contemporary Cuban and Latin American history, culture, and society, with a particular focus on armed struggle.

Leidy Casimiro Rodríguez is professor at the Universidad de Sancti Spíritus, Cuba, with more than fifteen years of experience in research and practice in the field of agroecology and permaculture. Leidy is a member of the Casimiro farming family at Finca del Medio. She has authored several articles in national and international scientific journals, book chapters, and a book titled *Resiliencia socioecologica de fincas familiares en Cuba*. She holds a degree in economics from Universidad Central de las Villas, law degree from Universidad de Sancti Spíritus, and PhD in agroecology from Universidad de Antioquia, Colombia.

Lauren Collins is an honorary research fellow at the University of Nottingham. She was awarded a PhD in Hispanic and Latin American Studies from the University of Nottingham in 2018 for her thesis on the history of local government in post-1959 Cuba. She is co-founder and director of the Caribbean Research and Innovation Collaboration for Knowledge Exchange and Transfer (CRICKET), a U.K.-based community interest company, which was established to bring together Caribbean and U.K. researchers to work on practical solution-orientated research projects.

Margarita Fernández has, since 2015, been the executive director of the Caribbean Agroecology Institute (CAI), a small nonprofit that catalyzes knowledge exchange, builds capacity, and supports transitions to agroecological systems that are resilient to climate change in Cuba and the region. Prior to this, she worked for twenty years in a number of different roles as a farmer, researcher, and practitioner in transdisciplinary research and the design, implementation, monitoring, and evaluation of international cooperation projects in the global south including Cuba, Laos, Mexico, Nicaragua, and Guatemala. Margarita has over a dozen articles published in peer-reviewed journals, book chapters, and has co-edited several books. She holds a PhD in agroecology from the University of Vermont, a master's in social ecology from the Yale School of Forestry and Environmental Studies, and a bachelor's in biology from Tufts University.

Shawn H. E. Harmon is a research associate at the IWK Health Centre, research scholar at the Dalhousie Health Law Institute, and part-time faculty at the Schulich School of Law, Dalhousie University, Canada. He is a former reader in law at the University of Edinburgh, UK, and co-founder and former deputy director of the J. Kenyon Mason Institute in Medicine, Life Sciences, and Law. Broadly concerned with health law, human rights, and global health justice, Dr. Harmon's research has examined the legal and ethical governance of emerging technologies and a wide range of health practices, including life science innovation, informed consent, abortion, and immunization, looking at both international and domestic frameworks. He has recently completed reports for Canadian governments of different levels relating to data collection, use, and access, and to public health governance. He is a member of the Royal Society of Canada's Working Group on Vaccine Acceptance.

Jessica Hirtle's graduate training focused on disaster preparedness and risk reduction in the Caribbean, receiving a MA in international development studies from Dalhousie University. From 2018 to 2020, she was a key researcher on an international study exploring comparative disaster management, conducting research in Canada and Cuba. Her current research analyzes disaster response strategies in Cuba, with an emphasis on the impact of solidary.

Hal P. Klepak is professor emeritus of the Royal Military College in Kingston, Ontario, Canada. He has served as strategic analyst at the National Defence Headquarters in Ottawa and NATO Headquarters in Brussels, and as a consultant on international security matters to the Foreign Affairs department of the Canadian government, international organizations, and academic boards. He is the author of *Cuba's Military, 1990–2005: Revolutionary Soldiers in Counter-revolutionary Times* (2005), *Raúl Castro: estratega de la defensa revolucionaria de Cuba* (2010), and *Churchill Comes of Age: Cuba 1895* (2015).

Giraldo Martín Martín was the director of the Pasture and Forage Experimental Station Indio Hatuey at the University of Matanzas, Cuba, from 1997 to 2020. He is currently a researcher and director of the national project Biomas-Cuba. He has more than thirty publications in national journals and books. He received his doctorate in Agrarian Science from the University of Matanzas.

Guillermo Mesa Ridel is the head of the special situations in the public health department at the National School of Public Health (ENSAP). He is a member of the Cuban Ministry of Public Health's National Working Group on tackling climate change in the health sector. Since 2015, he has also been the coordinator of international courses in health emergencies in large populations (HELP) and the subsequent management of humanitarian assistance. He has published extensively on climate change and disasters, including their relationships to the health sector.

Valerie Miller is the director of the Cuba Oceans Program for Environmental Defense Fund (EDF). In this position, she cultivates alliances across a multinational group of government, academic, and civil society partners to help develop and implement fisheries management and marine conservation projects in Cuba that can benefit the wider Gulf of Mexico and Caribbean region. She has a dual-degree MSc in conservation leadership from Colorado State University and El Colegio de la Frontera Sur. Her research has been published widely, including in journals such as *MEDICC Review*, *Bulletin of Marine Science*, and *Frontiers in Marine Science*.

Luis L. Vázquez received his bachelor and doctorate degrees in agronomy and sciences. He is currently an affiliate of the Cuban Association of Agronomists and Foresters (ACTAF). He worked for thirty years as a lead researcher at the Plant Health Research Institute (INISAV). Luis is an invited professor at several universities in Latin America and collaborates on various international cooperation projects. He has more than 100 publications in national and international journals, as well as several books.

Helen Yaffe is a lecturer in economic and social history at the University of Glasgow and a visiting fellow at the Latin American and Caribbean Centre at the London School of Economics. Her research and publications focus on Cuban development. Since 1995, she has spent time living and researching in Cuba. Her doctoral thesis was adapted for publication as *Che Guevara: The Economics of Revolution* in 2009 and in 2020 she published *We Are Cuba: How A Revolutionary People Have Survived in a Post-Soviet World.* She is also co-author of *Youth Activism and Solidarity: The Non-Stop Picket Against Apartheid* published in 2017.

www.ingramcontent.com/pod-product-compliance
Ingram Content Group UK Ltd.
Pitfield, Milton Keynes, MK11 3LW, UK
UKHW041842031026
15587UKWH00002B/942

9 781793 651310